# MARKS OF COURAGE

Edited by

**Ashley Cunningham-Boothe**

KORVET
Publishing & Distribution

First published 1991

Specially produced by Cedric Chivers Ltd, Bristol
for
Korvet Publishing & Distribution
Korvet House, Highfield Terrace, Leamington Spa, Warks CV32 6ER

ISBN 0 9517750 0 6

Printed in Great Britain by
Hillman Printers (Frome) Ltd, Frome, Somerset
Bound by Cedric Chivers Ltd, Bristol

Also by

Ashley Cunningham-Boothe

"British Forces in the Korean War"

Jointly edited with Dr Peter Farrar

Published by The British Korean Veterans Association

The complete record of Honours, Decorations and Awards for Gallant and Distinguished Services presented to members of the British and Commonwealth Forces in the Korean Theatre of War 1950 - 53.

"It is courage, courage, courage, that raises
the blood of life to crimson splendour"

George Bernard Shaw

Dedicated to the men and women from United Kingdom of Great Britain and Northern Ireland and the Commonwealth whose Citations of Bravery and of Distinguished Conduct have been honoured for their services with United Nations Forces during the Korean War.

ORDER OF PRECEDENCE

Royal Navy

Royal Fleet Auxiliary

41 Independent Commando, Royal Marines

Royal Marines

Military:

Headquarters (Staff)

Royal Armoured Corps

Royal Regiment of Artillery

Corps of Royal Engineers

Royal Corps of Signals

The Regiments of Infantry

Royal Army Chaplains' Department

Royal Army Service Corps

Royal Army Medical Corps

Royal Army Ordnance Corps

Corps of Royal Electrical and Mechanical Engineers

Corps of Royal Military Police

Royal Army Pay Corps

Military Provost Staff Corps

Royal Army Education Corps

Royal Army Dental Corps

Intelligence Corps

Army Physical Training Corps

Army Catering Corps

Queen Alexandra's Royal Army Nursing Corps

Royal Air Force

The Commonwealth Countries:

Australia

Canada

New Zealand

India

South Africa

Hong Kong

## SHIPS OF THE ROYAL AND COMMONWEALTH NAVIES ENGAGED IN OPERATIONS IN KOREA WATERS 1950 - 53

### ROYAL NAVY:

| | | | |
|---|---|---|---|
| Alacrity | Cockade | Hart | St Brides Bay |
| Alert | Comus | Jamaica | Sparrow |
| Amethyst | Concord | Kenya | Telemachus |
| Belfast | Consort | Modeste | Theseus |
| Birmingham | Constance | Morecambe Bay | Tobruk |
| Black Swan | Cossack | Mounts Bay | Triumph |
| Cardigan Bay | Crane | Newcastle | Tyne |
| Ceylon | Crusader | Ocean | Unicorn |
| Charity | Glory | Opossum | Whitesand Bay |

### ROYAL FLEET AUXILIARIES:

| | | | | |
|---|---|---|---|---|
| Brown Ranger | Wave Baron | Wave Knight | Wave Premier | Wave Sovereign |
| Green Ranger | Wave Chief | Wave Laird | Wave Prince | |

### ROYAL CANADIAN NAVY:

| | | | |
|---|---|---|---|
| Athabaskan | Haida | Iroquois | Sioux |
| Cayuga | Huron | Nootka | |

### ROYAL AUSTRALIAN NAVY:

| | | | |
|---|---|---|---|
| Anzac | Condamine | Murchison | Sydney |
| Bataan | Culgoa | Shoalhaven | Warramunga |

### ROYAL NEW ZEALAND NAVY:

| | | | | |
|---|---|---|---|---|
| Hawea | Kaniere | Putaki | Taupo | Tutira |

### FLEET AIR ARM SQUADRONS:

800, 801, 802, 804, 807, 808, 810, 812, 817, 821, 825, 898.

BRITISH REGIMENTS OF INFANTRY IN THE KOREAN WAR

Order of precedence (as at the time of the Korean War and prior to re-organisation)

| | | |
|---|---|---|
| 1 | 1st of Foot - The Royal Scots | raised 1635 |
| 2 | 5th of Foot - The Royal Northumberland Fusiliers | raised 1674 |
| 3 | 7th of Foot - The Royal Fusiliers (City of London Regiment) | raised 1685 |
| 4 | 8th of Foot - The King's Regiment Liverpool | raised 1685 |
| 5 | 9th of Foot - The Royal Norfolk Regiment | raised 1685 |
| 6 | 17th of Foot - The Royal Leicestershire Regiment | raised 1688 |
| 7 | 25th of Foot - The King's Own Scottish Borderers | raised 1689 |
| 8 | 28th of Foot - The Gloucestershire Regiment | raised 1694 |
| 9 | 33rd of Foot - The Duke of Wellington's Regiment (West Riding) | raised 1702 |
| 10 | 41st of Foot - The Welch Regiment | raised 1719 |
| 11 | 42nd of Foot - The Black Watch (Royal Highland Regiment) | raised 1739 |
| 12 | 53rd of Foot - The King's Shropshire Light Infantry | raised 1755 |
| 13 | 57th of Foot - The Middlesex Regiment (Duke of Cambridge's Own) | raised 1755 |
| 14 | 68th of Foot - The Durham Light Infantry | raised 1758 |
| 15 | 83rd of Foot - The Royal Ulster Rifles | raised 1793 |
| 16 | 91st of Foot - The Argyll and Sutherland Highlanders (Princess Louise's) | raised 1794 |

# FOREWORD

by

Major General Peter A. Downward, CB, DSO, DFC,

President

The British Korean Veterans Association

This is a major reference work of the honours, decorations and awards for gallant and distinguished services awarded to British and Commonwealth Forces in the Korean Theatre of War (1950-53), and includes awards to those who served with the British Commonwealth Occupation Forces in Japan (BCOF), where these are linked with Korean War operations.

It is a truly definitive record of the tributes - MARKS OF COURAGE - that have been made in recognition of individual courage and of the wonderful human enterprise of those who fought and served with distinction. No such record has previously been published in such a disciplined and wholly comprehensive form.

Forty years have elapsed since the outbreak of the Korean War, during which time two generations have grown up, many of whom have little concept of what the war was about, where the actions were fought, and indeed why so many nations were involved. . Those events are now history, and the record of the many individual acts of gallantry in these pages has come not a moment too soon. These tributes now complement the many regimental histories, and accounts from ships' logs and air operations that have been written by other authors over recent years.

I should like to pay tribute to the immense effort that has been put into the research and preparation prior to publication; in spite of the editor's disabilities, which find their origins in the Korean War.

It will rate as an invaluable perspective to historians, to researchers and others with professional or collector's interest in awards for the Korean War, as well as for those veterans who served with the men and women whose names are recorded herein.

P.A. Downward

# INTRODUCTION

When the Republic of Korea was invaded by communist forces from the north on 25th June 1950, the scene became set for one of the most important and bloody conflicts of the twentieth century. Without delay, British and Commonwealth governments declared their support for the United Nations Security Council resolutions calling for assistance to resist aggression.

Generally considered to have been a major conflict between two great powers - the United States of America and China - the contributions made by United Nations member countries, however, were considerable. None more so than by British and Commonwealth countries, elements of which fought in all the major battles from the outbreak from the Pusan Perimeter (the Naktong Line) in September 1950 until the signing of the Armistice on 27th July 1953; battles that are now legendary in the annals of military history.

In a war considered to be a Soldiers' War, the role of British and Commonwealth naval units could not be spectacular, by comparison, because they had neither an enemy fleet nor an air force to contend with. Yet their contribution was not only significant but on occasions quite audacious. The role of the Royal Marines, especially that of 41 Independent Commando during the fearsome running-battles withdrawal from the Chosin Reservoir in North Korea during December 1950 and January 1951, was outstanding and courageous.

Australian and Canadian infantry units served with great distinction, and support received from the Royal New Zealand Artillery and the paratroopers of the Indian Army Field Ambulance Unit are major talking points, even nowadays, among veterans of the Korean War.

This is their own record of Brave Conduct and Distinguished Service.

The decision to commit myself to this preponderant undertaking was straightforward and vested of all complication; I saw it almost as a duty to see that a proper record was established, once and for all; by way of paying tribute, if you like, to the magnificent record of those named herein; not least of these, friends and comrades from my own Regiment, the Royal Northumberland Fusiliers (The Fighting Fifth); not realizing at the time what an incredible undertaking it would prove to be.

Quite by accident, I confess, I discovered a major flaw in the Australian listings and this led me to renewed research. In turn, it resulted in a review of the Canadian listings. As a consequence, I spent four months revising and retyping the Commonwealth section to the book. I am now satisfied that MARKS OF COURAGE is an authoritative record of honours, decorations and awards for gallant and distinguished services in the Korean Theatre of War.

An early inclination was to exclude all awards that could not be substantiated in the London Gazette or Commonwealth equivalent publications, working on the premise that these are the only official government publications for the promulgation of such awards. However, such were the great numbers of awards that appeared in publications, such as those mentioned in the bibliography, unsupported by details of a Gazette entry, it was decided to be in the best interest of all to include them. These entries are recognisable by the absence of an issue number and an issue date of a Gazette notice.

Commonwealth awards are listed in the London Gazette, but regrettably too many were not. There is, in my view, a lamentable lack of detail in too many official listings, especially naval awards, in the London and Commonwealth Gazettes; as is particularly so in the case of Canadian awards.

More so than for the other Services, it was quite apparent that the Senior Service had been neglected by those responsible for the promulgation of naval awards, for all too many such awards were not credited with the Ship's title that the person served with at the time the award was made. This practice is regrettable. Imagine the chaos for historians, researchers, librarians and the many others with a curiosity for this subject, endeavouring to trace entries for the Military if regiments and corps were omitted from individual Gazette entries. Smith 021 and Smith 031 would be hopelessly irrelevant if not listed under specific unit identities. Thus, Smith, Royal Navy, is simply not good enough. Only a small number of British and Commonwealth Air Forces entries gazetted actually showed specific unit identity. I am particularly grateful, therefore, to the publications shown in the Bibliography, which did actually list them.

Almost without exception, each entry in MARKS OF COURAGE will display a Gazette issue number and date; highest rank held (including acting rank) at time of award; personal service serial number; surname and full Christian names, and unit. (Other than for its novelty value, there is no historical relevance in showing a rank of a Private as acting Staff Sergeant! Which seems to have been the case with one Commonwealth nurse.)

Those shown in the London Gazette as being attached to regiments other than their parent-regiments have been listed under the unit they served with at the time their awards were made. This cannot be the same for Corps personnel, who are listed under their specific Corps and, where shown in official listings, the name of the unit attached to.

Regrettably, in several instances, many Gazette entries do not show the name of a person's unit, so these have been listed under a Headquarters heading.

The order of precedence for the rankings and appointments of members of the Senior Service has been the cause of great consternation to the Editor, for even the Ministry of Defence were not able to advise. Many of the old-time rankings, and this is especially so in the case of Fleet Air Arm personnel, have been made redundant by the advent of new technology and modern armaments. So the naval sections of the book were the most difficult to get into good order. Earlier published listings skirted the problem by simply lumping naval awards in any fashion and without orderliness.

The Editor is grateful, therefore, to former naval friends in The British Korean Veterans Association, who did their 'Andrew' about the Peninsula during the Korean War, whom I have named in the Acknowledgements, for helping me to get rankings in somewhat good order. But if all is not entirely Bristol-fashion with this section of the awards, I shall plead ignorance and divert any critique to them - like a good 'Pongo'!

Those badged infantry or corps who were decorated for their Distinguished Flying Operations in Korea have been placed into a category of their own. To have 'lumped' them under a headquarters sub-heading, because the majority were badged with units that did not serve in Korea, was felt by the Editor to be quite wrong. Using editorial privilege, these are featured separately to recognize their fearless if somewhat madcap practice of flying flimsy, unarmed aircraft over enemy territory, as eyes of the Gunners.

41 Independent Commando, Royal Marines, too, have been noticeably separated from the main heading of Royal Marines. This has been done with deliberate purpose by the Editor, and is intended as a mark of great admiration for the superb professionalism of those who served with this unit during the Korean War, particularly those who were involved with our American allies during the continuous, running-battles withdrawal from the Chosin Reservoir in North Korea, December 1950 to January 1951.

In his book, Medals and Decorations, Ian Angus informs readers that: "The Civil War was the first campaign in which medals were awarded specifically for acts of bravery, rather than as a token of regard to leading generals.

At his Court held at Oxford in May 1643, King Charles I instituted medals 'to be delivered to wear on the breast of every man who shall be certified under the hands of their commanders-in-chief to have done faithful service in forlorn hope'. Two Forlorn Hope medals were produced, portraying the King and his son (later Charles II) respectively.

These medals, issued generally for bravery, were preceded by a gold medal which the King awarded to Robert Welch for saving the Royal Standard at the Battle of Edgehill on 23rd October, 1642, the first battle of the war. This unique award depicted the standard on the reverse and bore a Latin inscription paying tribute to Welch." End of quote.

In 1854, the Distinguished Conduct Medal, which supersedes as a bravery award the Meritorious Service Medal (instituted 1845), is considered to be one of the most important decorations for Warrant Officers, non-commissioned officers and enlisted men. The high regard for acts of bravery by the military establishment, when making judgement of such acts, is evident by the small number of awards of this highly-regarded bravery decoration.

The bar to the DCM was awarded twice during the Korean War. To Sgt Rowlinson, DCM of 2nd Bn The Royal Australian Regiment and to Cpl Major, DCM of 2nd Bn The Royal 22e Regiment of Canada. Rowlinson was the only man to be awarded the DCM and the bar in the Korean War.

The highest number of DCMs awarded within a single unit went to the 3rd Bn the Royal Australian Regiment; all being awarded within twelve months of each other, to Sgt William Josiah Rowlinson (including the bar), WO II Leonard Murray Opie, Sgt Edward John Morrison and to L/Cpl James Burnett.

The highest number of DCMs awarded in a British Regiment went to WO II (CSM) James Murdoch, Sgt John Moss and to Cpl Archibald Laidlaw of 1st Bn The King's Own Scottish Borderers. Regrettably, the award of the DCM to a Sgt J. Moss of this Regiment, shown in several other publications, is ficticious. The Editor is grateful to Korean War veteran, Major Kenneth U. Fraser of Headquarters The King's Own Scottish Borderers at Berwick-upon-Tweed, for laying this mythical award to rest.

There is a general misconception shared by many, and this is very evident amongst the ranks of my fellow veterans, that citations are indispensable to promulgated gallantry awards that are published in the London Gazette. But this is not so. Only those citations which relate to the highest awards for valour and distinguished conduct, and on rare occasions for awards of the Distinguished Service Cross, are published.

And there are actually some whose names appear within the pages of this book who have never seen their own citations.

One Citation only has been reproduced in MARKS OF COURAGE, so that those who may not have seen one, or are unaware of the nature of acts of gallantry and distinguished conduct will have this opportunity to look at one man's cited courage. The Citation is for the Distinguished Conduct Medal, which was awarded to Fusilier George Arthur Hodkinson of 1st Bn The Royal Fusiliers (The City of London Regiment), which reads as follows :-

DISTINGUISHED CONDUCT MEDAL
220508234 Fusilier George Arthur HODKINSON
East Surrey Regiment (attached Royal Fusiliers)
(City of London Regiment)

"Fusilier HODKINSON was the wireless operator of 10 Platoon, D Company, 1st Bn The Royal Fusiliers, and on the night 24/25 November, 1952, accompanied his platoon on OP PIMLICO. His platoon had orders to sweep the KIGONG-NI ridge and establish a firm base on the MOUND at the western end, so that 12 Platoon could pass through on its objective.

Leaving the minegap at 1840 hours, after an hour's approach march, he took part in the initial and minor operation, which lasted until the platoon was established on the MOUND at 2007 hours. During this time the platoon had made one contact, located a collection of a hundred or more enemy to their south and from 2007 hours till 0100 hours had been motared.

Shortly after 0100 hours and after the assaulting platoon, some four hundred yards to the east had been heavily fired upon with burp sub-machine guns, a party of enemy numbering up to twenty, came running up the MOUND to the top. They were brusquely forced back by small arms fire and grenades.

Then followed a series of quick probes by the enemy of up to eight strong and from various directions at fifteen minutes intervals. They came in creeping to throw concussion grenades and to fire their burp guns. After each probe the casualties mounted and those outlying moved nearer to the centre. Fusilier HODKINSON throughout this period transmitted clearly and without excitement the orders and reports of his platoon commander.

By 0200 hours, however, his platoon commander was wounded and was killed at approximately 0230 hours. From this moment Fusilier Hodkinson took full control of the direction of battle. The platoon had suffered casualties and all the NCOs except one had been killed or wounded. With a voice loud and clear and using perfect procedure, he continued on his own initiative to report at frequent intervals a concise account of his platoon's battle and to use his sten gun whenever he could.

To obtain excellent technical transmission, with fearless determination he stood on the highest point of the MOUND.

At 0224 hours some thirty of the enemy crept up and made a determined attack.The platoon commander was killed and Fusilier HODKINSON was wounded in the leg and face. He now sat upon the top and directed the artillery, tank and mortar fire with remarkable accuracy and effect, bringing them nearer the MOUND. Around him, inspired by his example, fought the survivors, back to back using such cover as there was, and with the wounded collected at their feet, Fusilier HODKINSON then urged all to save their few remaining grenades and meagre ammunition to use them in volleys and only when necessary.

At 0250 hours the platoon was ordered to return, but Fusilier HODKINSON replied sharply: 'We cannot carry our wounded and therefore we will not leave'.

An attempt was made to bandage his face but this offer was rudely refused as he said he could best operate the wireless without a field dressing on his face.

At this stage he sent a warning to Company Headquarters, reporting enemy movement towards them and showing concern for their safety.

At 0339 hours Fusilier HODKINSON praised the Artillery and added that their fire was holding the enemy off the MOUND.

Finally, a vicious sweep came over the position and burp guns and concussion grenades were all around. At 0422 hours, still on the MOUND, he turned to his set and said: 'This is it. They are coming in strength. We shall be over-run this time. Nothing can stop them now. They are coming up the hill. We are being over-run, we are being over-run...' And later six of twenty-two men, all wounded, returned in pairs. The outstanding devotion to duty and leadership displayed by Fusilier HODKINSON was an inspiration to those around him, and to those farther back who heard his calm and determined transmissions. After his platoon commander had been killed this man, showing skill far beyond the normal capabilities of a Fusilier continued to direct and correct with remarkable accuracy and effect, the defensive fire of the artillery, mortars and tanks. Throughout the action he showed a complete disregard for his own personal safety, and the fact that his platoon was not over-run earlier was entirely due to the continued close defensive fire directed by Fusilier HODKINSON.

This heroic stand of his platoon drew off the enemy from 12 Platoon and Company Headquarters and thus enabled the majority of these, when recalled, to return safely."

The Editor has spoken with many veterans of the Korean War; quite the majority being affiliated in membership to The British Korean Veterans Association, and their almost universal criticism of the records of Honours, Decorations and Awards for Gallant and Distinguished Services in the Korean Theatre of War, thus far published, do not reflect the true record. The omission of the names of those decorated with the Bronze Oak Leaf Emblem, have been Mentioned in Despatches or received the rarely-awarded King's or Queen's Commendations for Gallant and Distinguished Services, is a particular sore point with them. No other publications records these awards. To have not included them in MARKS OF COURAGE would have been an injustice of their magnificent devotion to duty and to their comrades. Every effort has been made by the Editor to make this chronicle of the history of THE FORGOTTEN WAR an authoritative one.

The following excerpt from the full quotation credited to William James (1842 - 1910), teacher, psychologist and philosopher, illuminates my feelings, precisely :-

"Only the very exceptional individuals push to their extremes."

Thus, having done so, those who were awarded Mention in Despatches and King's and Queen's Commendations deserve to be included in the record, which it is my honour and privilege to have compiled.

Ashley Cunningham-Boothe

Editor

L.to R: Major Scott Shore, Colonel Geoffrey Brennan, Mrs Gregson, Lieutenant-Colonel Hugh Fairgrieve and Major John Morgan. Behind: Colonel Georges Poupard, President of the French Korean War Veterans Association, on the occasion of the presentation of the Croix de Guerre, held at the French Embassy, London, 28th March, 1991.
Photograph by Peter Farrar.

Flanking the Editor of Mark of Courage, former regimental colleagues, Major David Sharp, BEM, on the left and Derek Kinne, GC, on the right. The occasion: the Regimental Reunion of The Royal Northumberland Fusiliers, Fenham Barracks, Newcastle-Upon-Tyne, St. George's Day, 1991, to commemorate the 40th anniversary of the now legendary Battle of the River Imjin, 1951.

Both David Sharp and Derek Kinne were captured during the battle and were subsequently decorated for their gallant and distinguished conduct as prisoners of war in North Korea. The Editor last saw David Sharp during the battle of the River Imjin, and Derek Kinne prior to the battle.
Photograph by Nick Urwin.

| London Gazette issue No. | London Gazette date | Personal Service number | Name | Christian names | Ship's title |
|---|---|---|---|---|---|
| **ROYAL NAVY** | | | | | |
| **KBE: KNIGHT COMMANDER OF THE BRITISH EMPIRE** | | | | | |
| **VICE ADMIRAL** | | | | | |
| 39138 | 02.02.51 | | ANDREWES, CB, CBE, DSO | William Gerrard | |
| **CBE: COMMANDER OF THE ORDER OF THE BRITISH EMPIRE** | | | | | |
| **REAR ADMIRALS** | | | | | |
| 40011 | 06.11.53 | | CLIFFORD, CB | Eric George Anderson | |
| 39660 | 03.10.52 | | SCOTT-MONCRIEFF, CB, DSO | Alan Kenneth | |
| **CAPTAINS** | | | | | |
| 39660 | 03.10.52 | | COLQUHOUN, DSO | Kenneth Stewert | HMS Glory |
| 39854 | 19.05.53 | | EVANS, DSO, DSC | Charles Leo Glandore | HMS Ocean |
| 39138 | 02.02.51 | | JAMES | Ughtred Henry Ransden | not listed |
| 40011 | 06.11.53 | | LEWIN, DSO, DSC | Edgar Duncan Goodenough | HMS Glory |
| **BAR TO DSO: DISTINGUISHED SERVICE ORDER** | | | | | |
| **CAPTAINS** | | | | | |
| 39725 | 23.12.52 | | DUCKWORTH, DSO, DSC | Auberon Charles Alan Campbell | HMS Belfast |
| 39138 | 02.02.51 | | SALTER, DSO, OBE | Jocelyn Stuart Cambridge | not listed |
| 39660 | 03.10.52 | | THRING, DSO | George Arthur | HMS Ceylon |
| **DSO: DISTINGUISHED SERVICE ORDER** | | | | | |
| **CAPTAINS** | | | | | |
| 39854 | 19.05.53 | | ADAIR, OBE | Walter Alexander | HMS Cossack |
| 39660 | 03.10.52 | | BEGG, DSC | Varyl Cargill | HMS Cossack |
| 39272 | 26.06.51 | | BOLT, DSC | Arthur Seymour | not listed |
| 39272 | 26.06.51 | | BROCK | Patrick Willet | not listed |
| 39547 | 23.05.52 | | BROWN, OBE, DSC | Walter Leslie Mortimer | not listed |
| 39725 | 23.12.52 | | COLERIDGE, DSC | Hugh Charles Bainbridge | HMS Cardigan Bay |
| 40109 | 19.02.54 | | DURANT, DSC | Bryan Cecil | HMS Cardigan Bay |
| 40011 | 06.11.53 | | GREENING, DSC | Charles Woollven | HMS Birmingham |
| 39547 | 23.05.52 | | LLOYD-DAVIES, DSC | Cromwell Felix Justin | not listed |
| 39854 | 19.05.53 | | MARSH | Richard Lovatt Haslam | HMS Crane |
| 39547 | 23.05.52 | | NORFOLK | George Anthony Francis | not listed |
| 40011 | 06.11.53 | | RUTHERFORD | William Francis Henry Crawford | HMS Newcastle |

| London Gazette issue No. | London Gazette date | Personal Service number | Name | Christian names | Ship's title |
|---|---|---|---|---|---|
| **DSO: DISTINGUISHED SERVICE ORDER (continued)** | | | | | |
| **COMMANDERS** | | | | | |
| 39854 | 19.05.53 | | ROBERTS | Cedric Kenelm | HMS Ocean |
| 39660 | 03.10.52 | | SWANTON, DSC | Francis Alan | HMS Glory |
| **LIEUTENANT-COMMANDERS** | | | | | |
| 39660 | 03.10.52 | | HALL, DSC | Sidney James | HMS Glory |
| 39547 | 23.05.52 | | KEIGHLY-PEACH | Peter Lindsey | not listed |
| **OBE: OFFICER OF THE ORDER OF THE BRITISH EMPIRE** | | | | | |
| **COMMANDERS** | | | | | |
| 39272 | 26.06.51 | | ANDREW | Malcolm Fraser | not listed |
| 40109 | 19.02.54 | | BAILEY | Oswald Nigel | HMS Ocean |
| 39870 | 01.06.53 | | LOMBARD-HOBSON | Samuel Richard Le Hunte | HMS Newcastle |
| 39725 | 23.12.52 | | MEARES, DSC | John Aubrey | HMS Ladybird |
| 39854 | 19.05.53 | | MURRAY, DSC | John Avens | not listed |
| 40109 | 19.02.54 | | PARKER, DSC | Wilfred John | HMS Comus |
| 40109 | 19.02.54 | | PERCY, DSC | Terence Gerard Vaughan | not listed |
| 40011 | 06.11.53 | | SLEIGH, DSO, DSC | James Wallace | not listed |
| 39547 | 23.05.52 | | WHITE | Robert | not listed |
| 40011 | 06.11.53 | (E) | WITHERS | Frank | HMS Cossack |
| **MBE: MEMBER OF THE ORDER OF THE BRITISH EMPIRE** | | | | | |
| **FOR COURAGE AND FORTITUDE OF A HIGH ORDER WHILST A PRISONER OF WAR IN NORTH KOREA** | | | | | |
| **LIEUTENANT (SP)** | | | | | |
| 40109 | 19.02.54 | | LANKFORD, RNVR | Dennis Augustus | |
| **MBE: MEMBER OF THE ORDER OF THE BRITISH EMPIRE** | | | | | |
| **LIEUTENANT-COMMANDERS (E)** | | | | | |
| 39547 | 23.05.52 | | BUDDEN | Allan Frederick | not listed |
| 39725 | 23.12.52 | | JOHNSON | Harry Marles | HMS Comus |
| 39854 | 19.05.53 | | LUBY | Leslie Philip | HMS Ocean |
| 39682 | 28.10.52 | | TUNSTALL | Robert Joseph | HMAS Sydney (on loan) |
| **LIEUTENANT-COMMANDERS** | | | | | |
| 39725 | 23.12.52 | | CARDALE | Peter John | HMS Belfast |
| 40011 | 06.11.53 | | DALRYMPLE-HAMILTON, DSC | North Edward Frederick | HMS Birmingham |
| 39138 | 02.02.51 | | DREYER | Raymond Garnier | not listed |
| 40109 | 19.02.54 | | HOWSE, DSC | Humphrey Derek | HMS Newcastle |
| 39584 | 19.05.53 | | LEA | Richard Arnold James | HMS Ocean |

| London Gazette issue No. | London Gazette date | Personal Service number | Name | Christian names | Ship's title |
|---|---|---|---|---|---|
| MBE: MEMBER OF THE ORDER OF THE BRITISH EMPIRE (continued) | | | | | |
| LIEUTENANT-COMMANDERS (continued) | | | | | |
| 39725 | 23.12.52 | | MILLN | James Richard | HMS Belfast |
| 40109 | 19.02.54 | | MOTH | Brian Coventon | HMS Cossack |
| 39725 | 23.12.52 | | OLLIVANT, DSC | Martin Spencer | HMS Cossack |
| 40011 | 06.11.53 | | PEARSON | Ian Francis | HMS Glory |
| 39725 | 23.12.52 | | POWER | Arthur MacKenzie | HMS Ceylon |
| 40109 | 19.02.54 | | SHEPPARD | David Richard | not listed |
| 39725 | 23.12.52 | | SMITH | William Thomas Rutherford | HMS Glory |
| 40011 | 06.11.53 | | VANN, DSC | Bernard Geoffrey | HMS Newcastle |
| LIEUTENANTS | | | | | |
| 39870 | 01.06.53 | | LIDDICOAT | Anthony | HMS Mounts Bay |
| 39547 | 23.05.52 | | ROSS | Mark | not listed |
| COMMISSIONED COMMUNICATIONS OFFICER, RN | | | | | |
| 39138 | 02.02.51 | | MR. SYMONS, BEM | Arthur Frederick | not listed |
| 2ND BAR TO DSC: DISTINGUISHED SERVICE CROSS | | | | | |
| COMMANDERS | | | | | |
| 39854 | 15.05.53 | | BAYLY, DSC | Patrick Uniacke | HMS Constance |
| 40011 | 06.11.53 | | GATEHOUSE, DSC | Richard | HMS Charity |
| LIEUTENANT-COMMANDERS | | | | | |
| 39854 | 15.05.53 | | GRAHAM, DSC | George Onslow | HMNZS Rotoiti (on loan) |
| 39138 | 02.02.51 | | LEE, DSC | Herbert Jack | not listed |
| BAR TO DSC: DISTINGUISHED SERVICE CROSS | | | | | |
| COMMANDERS | | | | | |
| 40011 | 06.11.53 | | CARTWRIGHT, DSC | John Cecil | HMS Opossum |
| 39725 | 23.12.52 | | FARNOL, DSC | James Jeffrey Edward | HMS Whitesand Bay |
| 39682 | 28.10.52 | | FELL, DSO, DSC | Michael Frampton | HMAS Sydney (on loan) |
| 40011 | 06.11.53 | | HAYES, DSC | Harry Stanley | HMS Cockade |
| 39725 | 23.12.52 | | KIMPTON, DSC | John Townshend | HMS Cockade |
| 39547 | 23.05.52 | | McLAUGHLIN, DSC | Ian David | not listed |
| 39660 | 03.10.52 | | ROWELL, OBE, DSC | Arthur Nichol | HMS Whitesand Bay |
| 39272 | 26.06.51 | | STOVIN-BRADFORD, DSC | Frederick | not listed |
| LIEUTENANT-COMMANDER | | | | | |
| 39272 | 26.06.51 | | SMITH, DSC | Patrick Gordon | not listed |
| LIEUTENANTS | | | | | |
| 39272 | 26.06.51 | | RUTHERFORD, DSC | Neil | not listed |
| 39725 | 23.12.52 | | SWANSTON, DSC | George Andrew | HMS Ceylon |

| London Gazette issue No. | London Gazette date | Personal Service number | Name | Christian names | Ship's title |
|---|---|---|---|---|---|
| DSC: DISTINGUISHED SERVICE CROSS | | | | | |
| COMMANDERS | | | | | |
| 40109 | 19.02.54 | | BLOOMER | Andrew William | HMS Ocean |
| 39725 | 23.12.52 | | BUTLER | Adrian Rothwell Lane | HMS Amethyst |
| 39725 | 23.12.52 | | CRADDOCK-HARTOPP, MBE | Kenneth Alston | HMNZS Taupo (on loan) |
| 39854 | 15.05.53 | | CRAIG-WALLER | Michael Waller Beaufort | HMS Whitesand Bay |
| 39660 | 03.10.52 | | ELDER, OBE | William George Caudlish | HMS St Brides Bay |
| 39660 | 03.10.52 | | FANSHAWE, OBE | Peter Evelyn | HMS Amethyst |
| 40011 | 06.11.53 | | HAMER, OBE | John Anthony Hodnot | HMS Morecambe Bay |
| 39660 | 03.10.52 | | HENLEY | John Arthur Cameron | HMS Charity |
| 39547 | 23.05.52 | | HENNESSY | Robert Angus Martin | not listed |
| 39547 | 23.05.52 | | LUARD | Robert Alexander Ingles | not listed |
| 39725 | 23.12.52 | | MILLS | Charles Piercy | HMS Concord |
| 39660 | 03.10.52 | | POLLOCK | Charles Edward | HMS Comus |
| 39725 | 23.12.52 | | ROWE | George Barton | HMS Consort |
| 39272 | 29.06.51 | | ROWELL, OBE | Arthur Nichol | HMS Whitesand Bay |
| LIEUTENANT-COMMANDERS | | | | | |
| 39660 | 03.10.52 | | BAILEY, OBE | John Savile | HMS Glory |
| 39660 | 03.10.52 | | BIRRELL | Maurice Andrew | HMS Glory |
| 39660 | 03.10.52 | | BLAKE | Reginald Howard Watson | HMS Glory |
| 39272 | 26.06.51 | | COY | Geoffrey Rolfe | not listed |
| 39272 | 26.06.51 | | EVANS | Bernard | not listed |
| 39870 | 01.06.53 | | GARDNER | James Robert Nigel | HMS Glory |
| 39725 | 23.12.52 | | LAWRENCE | Peter Reginald | HMS Black Swan |
| 40011 | 06.11.53 | | PEARCE | James Henry Silvester | HMS Glory |
| 39870 | 01.06.53 | | STUART | Peter Basil | HMS Glory |
| LIEUTENANTS (E) | | | | | |
| 39725 | 23.12.52 | | BARLOW | Peter | HMS Glory |
| 39547 | 23.05.52 | | JULIAN | Harry Graham | not listed |
| LIEUTENANTS | | | | | |
| 39854 | 15.05.53 | | CARMICHAEL | Peter | HMS Ocean |
| 39854 | 15.05.53 | | DAVIS | Peter Steel | HMS Ocean |
| 39660 | 03.10.52 | | HARVEY | John Gabriel Cavendish | HMS Glory |
| 39854 | 15.05.53 | | HAWKESWORTH | Richard Denison Rowan | HMS Ocean |
| 39725 | 23.12.52 | | HEFFORD | Frederick | HMS Glory |
| 39138 | 02.02.51 | | LAMB | Peter Melville | not listed |
| 39547 | 23.05.52 | | LAVENDER | Charles James | not listed |
| 40011 | 06.11.53 | | LEAHY | Alan John | HMS Glory |
| 40011 | 06.11.53 | | McCANDLESS | Robert John | HMS Glory |

| London Gazette issue No. | London Gazette date | Personal Service number | Name | Christian names | Ship's title |
|---|---|---|---|---|---|
| DSC: DISTINGUISHED SERVICE CROSS (continued) | | | | | |
| LIEUTENANTS (continued) | | | | | |
| 39725 | 23.12.52 | | McNAUGHTON | Douglas Arthur | HMS Glory |
| 40011 | 06.11.53 | | MILLETT | Paul | HMS Glory |
| 39547 | 23.05.52 | | NOBLE | William | not listed |
| 40011 | 06.11.53 | | SAMPLE | Geoffrey David Hutton | HMS Glory |
| 39547 | 23.05.52 | | WEBBER | Richard Scott Foster | RNZN (on loan) |
| COMMISSIONED PILOTS | | | | | |
| 39660 | 03.10.52 | | NEILSON | John Alexander | HMS Glory |
| 39125 | 23.12.52 | | PURNELL | Maurice Henry Charles | HMS Glory |
| COMMISSIONED GUNNER (TAS) | | | | | |
| 39272 | 29.06.51 | | MR. RANDALL | Harry William Nils | not listed |
| BAR TO DSM: DISTINGUISHED SERVICE MEDAL | | | | | |
| YEOMAN OF SIGNALS | | | | | |
| 39138 | 02.02.51 | D/JX159311 | CLARE, DSM | Douglas Robert James | HMS Cockade |
| DSM: DISTINGUISHED SERVICE MEDAL | | | | | |
| CHIEF PETTY OFFICERS | | | | | |
| 39547 | 23.05.52 | P/J104125 | LING | Walter | not listed |
| 39854 | 19.05.53 | C/J109738 | VENUS, BEM | Frederick Thomas | HMS Cossack |
| PILOT III | | | | | |
| 39547 | 23.05.52 | L/FX670576 | BOTTOMLEY | Frank Edward | not listed |
| 39272 | 29.06.51 | L/FX643770 | GRANT | Raymond Charles | not listed |
| 39547 | 23.05.52 | L/FX706230 | LINES | Peter Ronald | not listed |
| YEOMAN OF SIGNALS | | | | | |
| 39547 | 23.05.52 | P/JX154558 | DOUBLEDAY | William Oswald | not listed |
| PETTY OFFICERS | | | | | |
| 40011 | 06.11.53 | D/JX156530 | BIBBY | John | HMS Cockade |
| 39660 | 03.10.52 | P/JX156571 | CAINE | Martin Guy | HMS Ceylon |
| 39854 | 19.05.53 | P/JX777508 | LOCKHART | James Hamilton | HMS Charity |
| 39854 | 19.05.53 | C/JX760229 | RICHARDSON | Jack | HMS Mounts Bay |
| LEADING STOKER | | | | | |
| 39138 | 02.02.51 | D/SKX817966 | BANISTER (MECH.) | John William | not listed |
| LEADING SEAMAN | | | | | |
| 39547 | 23.05.52 | P/JX159680 | JESSOP | John | not listed |

| London Gazette issue No. | London Gazette date | Personal Service number | Name | Christian names | Ship's title |
|---|---|---|---|---|---|
| DSM: DISTINGUISHED SERVICE MEDAL (continued) | | | | | |
| AIRCREWMEN I | | | | | |
| 39547 | 23.05.52 | L/PX76496 | BEETON | Charles Frederick | not listed |
| 39138 | 02.02.51 | L/FX77509 | CREER | Kenneth Alwyn | not listed |
| 39854 | 19.05.53 | L/FX77282 | McCULLOCH | Charles Patrick | HMS Ocean |
| 39854 | 19.05.53 | L/FX704598 | POTTER | James Patrick | HMS Ocean |
| 39547 | 23.05.52 | L/FX97073 | SHIEL | Frederick Henry | not listed |
| ABLE SEAMEN | | | | | |
| 39725 | 23.12.52 | P/SSX845913 | EVANS | John | HMS Concord |
| 40011 | 06.11.53 | D/SSX864649 | GRAY | James Bowes | HMS St Brides Bay |
| 39547 | 23.05.52 | P/SSX867022 | ROBERTSON | James | not listed |
| 40011 | 06.11.53 | D/SSX841404 | TROY | Francis Patrick | HMS St Brides Bay |
| BEM:BRITISH EMPIRE MEDAL | | | | | |
| CHIEF ENGINE ROOM ARTIFICERS | | | | | |
| 39870 | 01.06.53 | P/MX51251 | CLARK | Douglas Percival | HMS Newcastle |
| 39854 | 19.05.53 | P/MX53674 | ELLIOTT | Gordon Alfred Thomas | HMS Comus |
| 39725 | 23.12.52 | D/MX49944 | HESLOP | William John Fearson | HMS Amethyst |
| 40011 | 06.11.53 | P/MX60652 | PLOWMAN | Frederick William | HMS Whitesand Bay |
| 39870 | 01.06.53 | D/MX55465 | SHERRIFF | John Norrish | HMS St Brides Bay |
| 39660 | 03.10.52 | C/MX57395 | TURP | George Jack | HMS Glory |
| 40109 | 19.02.54 | P/MX58504 | WEAIRE | Dennis Lloyd | HMS Modeste |
| CHIEF RADIO ELECTRICIAN | | | | | |
| 39854 | 19.05.53 | D/MX876580 | ELVE | Peter | HMS St Brides Bay |
| CHIEF ELECTRICAL ARTIFICER | | | | | |
| 39138 | 02.02.51 | D/MX60620 | RICHARDS | William Joseph | not listed |
| CHIEF ELECTRICIAN | | | | | |
| 40109 | 19.02.54 | P/MX766463 | FENSOME | Ian Andrew Oliver | HMS Comus |
| CHIEF PETTY OFFICER TELEGRAPHISTS | | | | | |
| 39138 | 02.02.51 | C/JX137115 | ELPHICK | Edwin George | not listed |
| 39725 | 23.12.52 | D/JX162140 | McCULLOUGH | Edward Walter | HMS Cardigan Bay |
| 39138 | 02.02.51 | C/JX132657 | MITCHELL | Ronald Frederick August | not listed |
| 39138 | 02.02.51 | D/JX138416 | ROCKSTRO | Herbert Stanley Tostevin | not listed |
| CHIEF YOMEN OF SIGNALS | | | | | |
| 40011 | 06.11.53 | P/JX142197 | ANDRESON | Jens William | HMS Newcastle |
| 40109 | 19.02.54 | C/JX142781 | NEALE | Percy Arthur | HMS Mounts Bay |
| 39138 | 02.02.51 | D/JX136590 | ROBINSON | James William | not listed |
| 39725 | 23.12.52 | D/JX132572 | STEPHENS, DSM | Jack | HMS Ocean |

| London Gazette issue No. | London Gazette date | Personal Service number | Name | Christian names | Ship's title |
|---|---|---|---|---|---|
| BEM: BRITISH EMPIRE MEDAL (continued) | | | | | |
| CHIEF ORDNANCE ARTIFICERS | | | | | |
| 39547 | 23.05.52 | C/MX46995 | BUTLER | Howard George | not listed |
| 39547 | 23.05.52 | C/MX47666 | DENDY | Leslie Charles | not listed |
| 40011 | 06.11.53 | P/MX54621 | HAZEL, DSM | Henry Frank | HMS Comus |
| CHIEF MECHANICIANS | | | | | |
| 39870 | 01.06.53 | C/KX93449 | CHRISTIE | George Donald | HMS Birmingham |
| 39272 | 29.06.51 | C/KX81635 | LAWRENCE | Albert Samuel | not listed |
| 39272 | 29.06.51 | P/KX84335 | MADDEN | Leslie | not listed |
| CHIEF AIRCRAFT ARTIFICERS | | | | | |
| 39660 | 03.10.52 | L/FX75010 | FOOKS | Arthur Charles | HMS Glory |
| 39138 | 02.02.51 | L/FX75200 | TURNER | Christopher | not listed |
| CHIEF PETTY OFFICER AIRMAN (AHI) | | | | | |
| 39272 | 26.06.51 | L/FX670453 | SHORT | George | not listed |
| CHIEF AIRMAN FITTERS (E) | | | | | |
| 40109 | 19.02.54 | L/FX80773 | BRIGHTON | Ronald | HMS Ocean |
| 40011 | 06.11.53 | L/FX90051 | CATER | Ronald | HMS Glory |
| CHIEF PETTY OFFICERS | | | | | |
| 39660 | 03.10.52 | C/JX131037 | COE | Reginald John | HMS Belfast |
| 39870 | 01.06.53 | P/JX153991 | JONES | Donald Herbert Townley | HMS Comus |
| 40011 | 06.11.53 | P/JX152929 | McCANN | Joseph | HMS Newcastle |
| 39854 | 19.05.53 | D/JX140882 | MARKS | Charles Edward Atherston | HMS Consort |
| 39272 | 29.06.51 | C/JX146128 | MOYES, DSM | Thomas Archibald Dunstan | not listed |
| 40011 | 06.11.53 | C/JX149076 | SEYMOUR | Robert | HMS Cossack |
| MASTER AT ARMS | | | | | |
| 40011 | 06.11.53 | C/MX96304 | CAMPBELL | William John | HMS Birmingham |
| PETTY OFFICER TELEGRAPHISTS | | | | | |
| 39725 | 23.12.52 | P/JX163506 | PAY | Maurice Eric | HMS Ladybird |
| 39725 | 23.12.52 | P/JX246348 | POMEROY | Ronald Leslie George | HMS Ladybird |
| 39547 | 23.05.52 | D/JX427888 | ROPER | Thomas | not listed |
| 40011 | 06.11.53 | D/JX371542 | WALLACE | William Alec | HMS Consort |
| PETTY OFFICER AIRMAN FITTER (E) | | | | | |
| 39870 | 01.06.53 | L/SFX816580 | LOCKHART | Edward | HMS Unicorn |
| PETTY OFFICER (TDI) | | | | | |
| 39272 | 26.06.51 | P/JX157488 | ROWLAND | Cecil George | not listed |
| PETTY OFFICERS | | | | | |
| 39547 | 23.05.52 | D/JX153073 | FELTHAM | William James | not listed |
| 39854 | 19.05.53 | P/JX583976 | SEAMAN | Douglas | HMS Whitesand Bay |
| LEADING SICK BERTH ATTENDANT | | | | | |
| 39138 | 02.02.51 | C/MX786882 | LING | John William | not listed |

| London Gazette issue No. | London Gazette date | Personal Service number | Name | Christian names | Ship's title |
|---|---|---|---|---|---|
| BEM: BRITISH EMPIRE MEDAL (continued) | | | | | |
| LEADING SEAMAN | | | | | |
| 39725 | 23.12.52 | C/JX795648 | SMITH | Colin Alan Pilcher | HMS Mounts Bays |
| LEADING AIRCRAFT MECHANIC (E) | | | | | |
| 39660 | 03.10.52 | L/FX762191 | JONES | Peter Ernest | HMS Glory |
| MECHANICIAN FIRST CLASS | | | | | |
| 39272 | 26.06.51 | D/KX96135 | TOOLEY | David | not listed |
| STOKER MECHANIC | | | | | |
| 39660 | 03.10.52 | D/SKX841347 | MORTIMORE | Walter John | HMS Cardigan Bay |
| RADIO ELECTRICAL ARTIFICER (4) | | | | | |
| 39547 | 23.05.52 | C/MX708107 | ROCHE | Francis Gilbert | not listed |
| AIRCRAFT ARTIFICERS | | | | | |
| 39660 | 03.10.52 | L/FX87492 | ABBOTT | John Stanley | HMS Glory |
| 40011 | 03.11.53 | L/FX87512 | CHISHOLM | Jerrold Peter | HMS Glory |
| 39725 | 23.12.52 | L/FX89785 | GREEN | Leslie | HMS Glory |
| ELECTRICIAN (AIR) | | | | | |
| 39725 | 23.12.52 | L/FX513084 | BRICE | Herbert Geoffrey | HMS Glory |
| MENTION IN DESPATCHES (FOR GALLANT AND DISTINGUISHED SERVICES) - POSTHUMOUS | | | | | |
| BRONZE OAK LEAF EMBLEM | | | | | |
| LIEUTENANT COMMANDERS | | | | | |
| | | | DICK, DSC | Donald Arthur | HMS Ocean |
| 39138 | 02.02.51 | | MacLACHLAN | Ian Murray | not listed |
| LIEUTENANTS | | | | | |
| 39547 | 23.05.52 | | COOLES | Geoffrey Hammond | not listed |
| 39547 | 23.05.52 | | WILLIAMS | Robert | not listed |
| SURGEON LIEUTENANT | | | | | |
| 39230 | 18.05.51 | | KNOCK, MB, BS | Douglas Alexander | not listed |
| PETTY OFFICER | | | | | |
| 39230 | 18.05.51 | C/JX166456 | TATE | John Arnison | not listed |
| ABLE SEAMAN | | | | | |
| 39660 | 03.10.52 | D/SSX836037 | SKELTON | Clifford | HMS Cockade |
| MENTION IN DESPATCHES (FOR GALLANT AND DISTINGUISHED SERVICES) | | | | | |
| BRONZE OAK LEAF EMBLEM | | | | | |
| REAR ADMIRAL | | | | | |
| 39725 | 23.12.52 | | SCOTT-MONCRIEFF,CB,CBE,DSO | Alan Kennett | |
| COMMODORE | | | | | |
| 39660 | 03.10.52 | | ST CLAIR-FORD, Bt.,DSO | Sir Aubrey | HMS Belfast |

| London Gazette issue No. | London Gazette date | Persona; Service number | Name | Christian names | Ship's title |
|---|---|---|---|---|---|
| MENTION IN DESPATCHES (FOR GALLANT AND DISTINGUISHED SERVICES)(continued) | | | | | |
| CAPTAINS | | | | | |
| 39272 | 26.06.52 | | BEGG, DSC | Varyl Cargill | HMS Cossack |
| 39138 | 30.01.51 | | BROCK | Patrick Willet | not listed |
| 39272 | 26.06.52 | | BROWN, OBE, DSC | Walter Leslie Mortimer | not listed |
| 39660 | 03.10.52 | | FREWEN | John Byng | HMS Mounts Bay |
| 39272 | 26.06.52 | | HOPKINS, DSO, DSC | Frank, Henry Edward | not listed |
| 39138 | 02.02.51 | | JAY, DSO, DSC | Alan David Hastings | not listed |
| 39547 | 23.05.52 | | LARKEN | Francis Wyatt Rawson | not listed |
| 39854 | 19.05.53 | | LEWIS | Arthur Francis Patrick | HMS Mounts Bay |
| 40109 | 19.02.54 | | LOGAN | Brian Ewen Weldon | HMS Ocean |
| 39854 | 19.05.53 | | LUCE, DSO, OBE | John David | HMS Birmingham |
| 39660 | 03.10.52 | | PODGER | Theodore Edward | HMS Kenya |
| 39138 | 02.02.51 | | ST CLAIR-FORD,Bt, DSO | Sir Aubrey | HMS Belfast |
| 39660 | 03.10.52 | | THOMPSON | John Yelverton | HMS Unicorn |
| 39139 | 02.02.51 | | TORLESSE, DSO | Arthur David | not listed |
| 39138 | 02.02.51 | | WHITE, DSO | Richard Taylor | not listed |
| SURGEON-CAPTAIN | | | | | |
| 39139 | 02.02.51 | | LYNAGH, MB, B.Ch. | Thomas Barnard | not listed |
| COMMANDERS | | | | | |
| 39660 | 03.10.52 | | ALEXANDER, DSO, DSC | Robert Love | HMS Glory |
| 39138 | 02.02.51 | | ANDREW | Malcolm Fraser | not listed |
| 40109 | 19.02.54 | | BELLARS, MBE | Anthony Gerald William | not listed |
| 39660 | 03.10.52 | | BEST | Thomas William | not listed |
| 39660 | 03.10.52 | | BLACKHAM | Joseph Leslie | HMS Belfast |
| 39138 | 02.02.51 | | BRUCE | Merlin | not listed |
| 39272 | 29.06.51 | | CARVER, DSC | Edmund Squarey | not listed |
| 39272 | 29.06.51 | | COMPSTON | Peter Maxwell | not listed |
| 40109 | 19.02.54 | | CREE, DSC | Charles Eustace Audley | HMS Birmingham |
| 40109 | 19.02.54 | | DIXON, MBE, DSC | John Peel | not listed |
| 39547 | 23.05.52 | | FANSHAW, OBE | Peter Evelyn | HMS Amethyst |
| 39138 | 30.01.51 | | HAMILTON | James Kennett | not listed |
| 39336 | 19.09.51 | | HERRICK, DSC | Laurence Edward | RNZN (on loan) |
| 39682 | 28.10.52 | | KIGGELL, DSC | L.J. | HMAS Sydney (on loan) |
| 39272 | 29.06.51 | | MUNN, DSO, OBE | William James | not listed |
| 39272 | 29.06.51 | | SEALE, DSC | Allan George Luscombe | not listed |
| 39660 | 03.10.52 | | SHAW, DSC | Terence Waters Brown | HMS Ceylon |
| 40001 | 06.11.53 | | TEALE | William John MacDonald | HMS Newcastle |
| 39854 | 19.05.53 | | WESTERN | James Guy Trench | HMS St Brides Bay |

| London Gazette issue No. | London Gazette date | Personal Service number | Name | Christian names | Ship's title |
|---|---|---|---|---|---|
| MENTION IN DESPATCHES (FOR GALLANT AND DISTINGUISHED SERVICES)(continued) | | | | | |
| COMMANDER (E) | | | | | |
| 39272 | 29.06.51 | | CLARK | Jack Armstrong Fenn | not listed |
| 39138 | 02.02.51 | | SMITH, AMI Mech.E, MI Mar.E | Alan Funge | not listed |
| 39272 | 29.06.51 | | THOMPSON | George Alan | not listed |
| 39139 | 02.02.51 | | TROLLOPE | Peter Francis | not listed |
| 39725 | 23.12.52 | | TYNDALE-BISCOE | Edward Rupert | HMS Belfast |
| COMMANDER (L) | | | | | |
| 39870 | 01.06.53 | | GRIFFITH | Anthony George Bruce | HMS Cossack |
| COMMANDER (S) | | | | | |
| 39725 | 23.12.52 | | FIELD, DSC | William Hugh | HMS Glory |
| 40011 | 06.11.53 | | HOWELL-DAVIES | Mervyn David | HMS Newcastle |
| 39138 | 02.02.51 | | STANNING, DSO | Geoffrey Heaton | not listed |
| LIEUTENANT-COMMANDERS | | | | | |
| 39138 | 30.01.51 | | ALDRIDGE, MBE | John Luttrell | not listed |
| 40011 | 06.11.53 | | ALSTON | Jack Rowland | HMS Cossack |
| 39854 | 19.05.53 | | BOWEN | Anthony | HMS Consort |
| 40011 | 06.11.53 | | BROWN | Donald George | HMS Mounts Bay |
| | | | DICK, DSC | D.A. | HMS Ocean |
| 39272 | 29.06.51 | | DODDS | Anthony Kirkwood | not listed |
| 40011 | 06.11.53 | | ELLIS | Arthur Woodhouse | HMS Crane |
| 39272 | 29.06.51 | | GORE-LANGTON | Alaric Hubert St George | not listed |
| 39660 | 03.10.52 | | HAMILTON, DSC | Alastair Gaven | HMS Concord |
| 39138 | 02.02.51 | | HANDLEY | Thomas Dennis | not listed |
| 40011 | 06.11.53 | | HARWOOD | Cecil Michael | HMS Charity |
| 39138 | 02.02.51 | | HENESSY | Robert Angus Martin | not listed |
| 39326 | 25.05.51 | | HOARE | Peter James Hill | RNZN (on loan) |
| 40011 | 06.11.53 | | JOY | Charles Henry Francis | not listed |
| 40011 | 06.11.53 | | McMILLAN | Colin Lovat Farquhar | HMS Morecambe Bay |
| 39870 | 01.06.53 | | MILES | Peter Tremayne | HMS Crane |
| 39660 | 03.10.52 | | OLLIVANT, DSC | Martin Spencer | HMS Cossack |
| 39272 | 26.06.51 | | PATTISSON, DSC | Kenneth Stuart | not listed |
| 39870 | 01.06.53 | | PHILLIPS, DSC | Seely | HMS Birmingham |
| 39547 | 23.05.52 | | RIDGEWAY | Thomas Graeme | not listed |
| 40109 | 19.02.54 | | ROBERTS | Arthur Douglas | HMS Crane |
| 39725 | 23.12.52 | | ROE | Richard Edward | HMS Cardigan Bay |
| 39725 | 23.12.52 | | SPADEMAN | Philip Reginald | HMS Glory |
| 39272 | 26.06.51 | | SYMS | Dudley Lester | not listed |
| 39725 | 23.12.52 | | TURNER | Norman Richard | HMS Alacrity |
| 39272 | 29.06.51 | | WORTH, DSC | Peter Reginald Glenholme | not listed |
| 40011 | 06.11.53 | | YOUNG | Robert | HMS Opossum |

| London Gazette issue No. | London Gazette date | Personal Service number | Name | Christian names | Ship's title |
|---|---|---|---|---|---|
| MENTION IN DESPATCHES (FOR GALLANT AND DISTINGUISHED SERVICES)(continued) | | | | | |
| BRONZE OAK LEAF EMBLEM | | | | | |
| LIEUTENANT-COMMANDER (E) | | | | | |
| 39725 | 23.12.52 | | ALEXANDER | David Leslie | HMS Crane |
| 39660 | 03.10.52 | | BURDEN | Walter George | HMS Charity |
| 40109 | 19.02.54 | | HARCUS | Ronald Albert | HMS Ocean |
| 40011 | 06.11.53 | | MORRIS | Anthony Charles Temple | HMS Birmingham |
| 39660 | 03.10.52 | | PEARSON | Ian Francis | HMS Glory |
| 39854 | 19.05.53 | | REID | Christopher Gordon Lestock | HMS Cardigan Bay |
| 39138 | 02.02.51 | | RICHARDS, AMI Mech.E | Michael | not listed |
| 40011 | 06.11.53 | | SATOW | Derek Graham | HMS Newcastle |
| 39854 | 19.05.53 | | SPENCER | Philip Edwin | HMS Constance |
| 39870 | 01.06.53 | | SULLIVAN | Eric John | HMS Concord |
| LIEUTENANT-COMMANDER (L) | | | | | |
| 39138 | 02.02.51 | | WYKEHAM-MARTIN | Herbert Roger | not listed |
| LIEUTENANT-COMMANDER (S) | | | | | |
| 39547 | 23.05.52 | | BUTLIN | Charles Edward | not listed |
| SURGEON LIEUTENANT-COMMANDER | | | | | |
| 39660 | 03.10.52 | | RITCHIE, MB, B.Ch. | James Simpson | HMS Morecambe Bay |
| ROYAL NAVY CHAPLAINS' DEPARTMENT | | | | | |
| 39660 | 03.10.52 | | THE REV FRY | Hugh Selwyn | HMS Ceylon (Chaplain) |
| LIEUTENANTS | | | | | |
| 39547 | 23.05.52 | | AUSTIN | Anthony John | not listed |
| 39660 | 03.10.52 | | BINNEY | Thomas Victor Giles | HMS Glory |
| 39854 | 19.05.53 | | BROWN | Michael Lawrence | HMS Ocean |
| 39660 | 03.10.52 | | BUTLER | Michael William | HMS Cardigan Bay |
| 39854 | 19.05.53 | | COOPER | William James | HMS Ocean |
| 39547 | 23.05.52 | | DAVIS | David John Holmes | not listed |
| 39854 | 19.05.53 | | DUNNE | Arthur Gordon | HMS Morecambe Bay |
| 39725 | 23.12.52 | | EDNEY | Walter Percy | HMS Constance |
| 39725 | 23.12.52 | | FIELDHOUSE | Derek Frederick | HMS Glory |
| 39272 | 29.06.51 | | FORD | Albert | not listed |
| 40011 | 06.11.53 | | FOSTER | Michael John Neville | HMS St Brides Bay |
| 39682 | 28.10.52 | | GENGE | Edward Thomas | HMAS Sydney (on loan) |
| 39870 | 01.06.53 | | GEORGE | Malcolm Atchison | HMS Opossum |
| 39854 | 19.05.53 | | HALLAM | Robert Henry | HMS Ocean |
| 39547 | 23.05.52 | | HIRST | David Charles | not listed |
| 39725 | 23.12.52 | | HOOPER | Anthony Desmond | HMS Glory |
| 39870 | 01.06.53 | | LEA | Edward Morris | HMS Mounts Bay |

| London Gazette issue No. | London Gazette date | Personal Service number | Name | Christian names | Ship's title |
|---|---|---|---|---|---|
| MENTION IN DESPATCHES (FOR GALLANT AND DISTINGUISHED SERVICES)(continued) | | | | | |
| BRONZE OAK LEAF EMBLEM | | | | | |
| LIEUTENANTS (continued) | | | | | |
| 40011 | 06.11.53 | | LLOYD | Geoffrey Charles | not listed |
| 39660 | 03.10.52 | | LUCAS | Kenneth Prior | HMS St Brides Bay |
| 39854 | 19.05.53 | | McKEOWN | David Thomas | HMS Ocean |
| 39725 | 23.12.52 | | MacKILLIGAN | William Hector | HMS Ceylon |
| 40011 | 06.11.53 | | MANN | Michael John | HMS Cardigan Bay |
| 39725 | 23.12.52 | | MARTINEAU | Philip | HMS Charity |
| 39660 | 03.10.52 | | MASLEN | Peter | HMS Amethyst |
| 39660 | 03.10.52 | | MILLS | John Peter | HMS Mounts Bay |
| 39138 | 02.02.51 | | MOSELEY | Charles Sheridan | not listed |
| 39854 | 19.05.53 | | PENISTON-BIRD | Norman Edmund | HMS Ocean |
| 39660 | 03.10.52 | | READ | Richard George Howard | HMS Comus |
| 39547 | 23.05.52 | | TEMPLE | Gerald Young | not listed |
| 39660 | 03.10.52 | | TRELAWNEY | Robin Christopher Beaumont | HMS Glory |
| 39854 | 19.05.53 | | TRELOAR | John Lewis | HMS Ocean |
| 39272 | 29.06.51 | | TRETHOWEN | Stanley Charles | not listed |
| 39272 | 29.06.51 | | WAKE-WALKER | Cedric Collingwood | not listed |
| 39854 | 19.05.53 | | WATKINSON | Peter | HMS Ocean |
| 39725 | 23.12.52 | | WHITAKER | Kenneth | HMS Glory |
| 39547 | 23.05.52 | | WILLIAMS | Neville Richard | HMS Theseus |
| 39272 | 29.06.51 | | WINTERTON | David Willoughby | not listed |
| 39725 | 23.12.52 | | WILSON | Roi Edgerton | HMS Glory |
| LIEUTENANT (E) | | | | | |
| 39660 | 03.10.52 | | FOSTER | Robin Beadon Lisle | HMS Unicorn |
| 39854 | 19.05.53 | | GARNER | Alan Leslie | HMS Whitesand Bay |
| LIEUTENANT (L) | | | | | |
| 39725 | 23.12.52 | | HASTON | Campbell Fraser | HMS Comus |
| 39272 | 29.06.51 | | HOCKEN | Melville Ruan | not listed |
| 39725 | 23.12.52 | | TRAVERS | Frank Guy | HMS Crane |
| SURGEON-LIEUTENANT | | | | | |
| 39725 | 23.12.52 | | ROWAN, MB, B.Ch., RNVR | Robert Anthony | HMS Belfast |
| SENIOR COMMISSIONED ENGINEERS | | | | | |
| 39854 | 19.05.53 | | BURT | Robert Case | HMS St Brides Bay |
| 40109 | 19.02.54 | | SMITH, DSM | Eric Nelson | HMS Morecambe Bay |
| SENIOR COMMISSIONED ELECTRICAL OFFICER (R) | | | | | |
| 39854 | 19.05.53 | | McBRIDE | Edwin Charles | HMS Belfast |
| SENIOR COMMISSIONED ORDNANCE ENGINEER | | | | | |
| 39660 | 03.10.52 | | ELLIOTT | Victor William Josiah | HMS Belfast |

| London Gazette issue No. | London Gazette date | Personal Service number | Name | Christian names | Ship's title |
|---|---|---|---|---|---|
| MENTION IN DESPATCHES (FOR GALLANT AND DISTINGUISHED SERVICES)(continued) | | | | | |
| BRONZE OAK LEAF EMBLEM | | | | | |
| SENIOR COMMISSIONED COMMUNICATIONS OFFICER | | | | | |
| 40109 | 19.02.54 | | KENNY | Frank Walter | HMS Birmingham |
| SENIOR COMMISSIONED GUNNER | | | | | |
| 39725 | 23.12.52 | | POTIER, DSM | George Raymond | HMS Belfast |
| COMMISSIONED ELECTRICAL OFFICER (R) | | | | | |
| 40011 | 06.11.53 | | FOWLER | James Hastings Stewart | HMS Birmingham |
| COMMISSIONED COMMUNICATIONS OFFICER | | | | | |
| 39725 | 23.12.52 | | MORRIS, DSM | Fred | HMS Cossack |
| COMMISSIONED GUNNER | | | | | |
| 40011 | 06.11.53 | | NUTE | Eric William | HMS Whitesand Bay |
| COMMISSIONED PILOTS | | | | | |
| 39547 | 23.05.52 | | BAILEY | Francis Dominic | not listed |
| 39725 | 23.12.52 | | DARLINGTON | Michael Ian | HMS Glory |
| 39660 | 03.10.52 | | HEFFORD | Frederick | HMS Glory |
| 39547 | 23.05.52 | | MacKENZIE | Ian | not listed |
| COMMISSIONED AIR ENGINEER | | | | | |
| 39870 | 01.06.53 | | JONES, BEM | Sydney | HMS Glory |
| COMMISSIONED AIR ENGINEER (ORDNANCE) | | | | | |
| 39660 | 03.10.52 | | FREELAND | James Henry | HMS Glory |
| COMMISSIONED AIRMAN | | | | | |
| 39547 | 23.05.52 | | MOTTRAM | Dennis Raymond | not listed |
| CHIEF YEOMAN OF SIGNALS | | | | | |
| 39272 | 29.06.51 | C/JX134676 | CARTER | Herbert Winston | not listed |
| 39272 | 29.06.51 | C/JX132880 | SMITH | Reginald Joseph | not listed |
| 40011 | 26.11.52 | D/JX134347 | WILSON | Claud Frank Thomas | HMS Crane |
| CHIEF PETTY OFFICER WRITERS | | | | | |
| 39138 | 02.02.51 | P/MX58749 | EDWARDS | Donald William | not listed |
| 40011 | 06.11.52 | P/MX70051 | GENT | Charles Arthur | HMS Newcastle |
| 39138 | 02.02.51 | D/MX51105 | NICHOLLS | Stanley James | not listed |
| CHIEF SICK BERTH PETTY OFFICER | | | | | |
| 39138 | 02.02.51 | D/MX49103 | ELSTON | Charles Edward | not listed |
| CHIEF PETTY OFFICERS | | | | | |
| 39547 | 23.05.52 | D/MX50792 | BARNETT | Henry | not listed |
| 39138 | 02.02.51 | C/JX130622 | BROWN | Albert Ronald | not listed |
| 39272 | 29.06.51 | C/JX134863 | CARR | Maurice Milner | not listed |
| 40011 | 06.11.52 | D/JX151810 | GUY | Arthur William Charles | HMS Cardigan Bay |
| 40109 | 19.02.54 | P/JX147812 | JONES | William Arthur | HMS Whitesand Bay |
| 39138 | 02.02.51 | D/JX132464 | MEAR | William Frank | not listed |

| London Gazette issue No. | London Gazette date | Personal Service number | Name | Christian names | Ship's title |
|---|---|---|---|---|---|
| MENTION IN DESPATCHES (FOR GALLANT AND DISTINGUISHED SERVICES)(continued) | | | | | |
| BRONZE OAK LEAF EMBLEM | | | | | |
| CHIEF PETTY OFFICERS (continued) | | | | | |
| 39138 | 02.02.51 | C/JX129946 | MUFFETT | Francis | not listed |
| 39870 | 01.06.53 | P/JX131384 | SMITH | Albert | HMS Morecambe Bay |
| CHIEF PETTY OFFICER AIRMAN | | | | | |
| 39725 | 23.12.52 | L/FX670499 | SADLER | Albert | HMS Glory |
| CHIEF AIRMEN | | | | | |
| 39854 | 19.05.53 | L/FX842372 | DIXON | Alan | HMS Ocean |
| 39854 | 19.05.53 | L/FX77094 | REID | Stanley | HMS Ocean |
| 39854 | 19.05.53 | C/JX646398 | SMITH | John Martin | HMS Cossack |
| CHIEF PETTY OFFICER STOKER MECHANIC | | | | | |
| 40011 | 06.11.52 | D/KX84660 | BAZELY | Reginald James | HMS Sparrow |
| 39547 | 23.05.52 | C/KX80637 | BRINE | William Lionel | not listed |
| 39725 | 23.12.53 | C/KX81420 | FOWLE | Basil George Charles | HMS Belfast |
| 39725 | 23.12.53 | C/KX83533 | PARKES | Arthur Francis George | HMS Belfast |
| 40011 | 06.11.52 | C/KX88267 | PAWLEY | Dennis Alfred | HMS Birmingham |
| 40011 | 06.11.52 | P/KX88051 | PENFOLD | William Frederick Henry | HMS Newcastle |
| 39854 | 15.05.53 | C/KX83836 | PITTAM | Herbert James | HMS Belfast |
| 39547 | 23.05.52 | C/KX89016 | SNELLING | Kendall Douglas | not listed |
| 40011 | 06.11.52 | P/KX86244 | TRINDER | Robert George Ernest | HMS Concord |
| CHIEF ELECTRICAL ARTIFICERS | | | | | |
| 39725 | 23.12.52 | C/MX47363 | BATES | Maurice John | HMS Belfast |
| 40011 | 06.11.52 | P/MX49966 | BLACK | Rupert George | HMS Concord |
| 39138 | 02.02.51 | C/MX46516 | DACEY | George Robert | not listed |
| 39660 | 03.]0.52 | C/MX804420 | MARLING | Fredrick Charles | HMS Belfast |
| 40011 | 06.11.52 | P/MX61693 | TOON | Ernest Harold | HMS Newcastle |
| CHIEF RADIO ELECTRICAL ARTIFICER | | | | | |
| 39725 | 23.12.52 | P/MX801845 | PARRY | Roy Cowell | HMS Alacrity |
| 39272 | 29.06.51 | C/MX126303 | TAYLOR | Harold Raymond | not listed |
| CHIEF ENGINE ROOM ARTIFICERS | | | | | |
| 39854 | 19.05.53 | P/MX50009 | BLACKMAN | Victor | HMS Morecambe Bay |
| 39725 | 23.12.52 | C/MX62545 | BRETT | Alfred James | HMS Glory |
| 39272 | 29.06.51 | D/MX53002 | CHAPMAN | William Henry | not listed |
| 39139 | 02.02.51 | C/MX50960 | CHAPPELL | Thomas | not listed |
| 39725 | 23.12.52 | D/MX52937 | CLOUGH | Reginald | HMS Cardigan Bay |
| 39660 | 03.10.52 | C/MX50348 | CONHEENEY | Michael | HMS Glory |

| London Gazette issue No. | London Gazette date | Personal Service number | Name | Christian names | Ship's title |
|---|---|---|---|---|---|
| MENTION IN DESPATCHES (FOR GALLANT AND DISTINGUISHED SERVICES)(continued) | | | | | |
| BRONZE OAK LEAF EMBLEM | | | | | |
| 39725 | 23.12.52 | C/MX48230 | CURTCHER | Stephen William | HMS Cossack |
| 39854 | 19.05.53 | C/MX51832 | DOREY | Charles Edward | HMS Mounts Bay |
| 39660 | 03.10.52 | P/MX60776 | FEENEY | John Howard | HMS Comus |
| 39272 | 29.06.51 | C/MX48370 | HEWITT, DSM | William Henry | not listed |
| 39725 | 23.12.52 | P/MX58109 | HOLYOME | Charles Stephen | HMS Alacrity |
| 39725 | 23.12.52 | P/MX48916 | McNAUGHTON | James William | HMS Whitesand Bay |
| 39725 | 23.12.52 | D/MX61560 | PERRIN, BEM | Douglas | HMS Ocean |
| 40011 | 06.11.52 | C/MX60607 | RAILER | Dennis | HMS Birmingham |
| 40109 | 19.02.54 | D/MX60250 | ROWE | James Pooley | HMS Ocean |
| 39547 | 23.05.52 | P/MX61795 | ROXBOROUGH | Andrew Houston | not listed |
| 39725 | 23.12.52 | P/MX50502 | SELLICK, DSM | John Bernard | HMS Charity |
| 40011 | 06.11.52 | D/MX60519 | SLACK | John | HMS Cardigan Bay |
| 39272 | 29.06.51 | P/MX54955 | SMITH | Thomas Henry | not listed |
| 40011 | 06.11.52 | D/MX63035 | WILLIAMS | Albert | HMS Opossum |
| 39725 | 23.12.52 | C/MX56868 | WILLIAMS, BEM | David Lloyd | HMS Belfast |
| CHIEF RADIO ELECTRICIAN | | | | | |
| 39547 | 23.05.52 | P/MX745789 | DAVIDSON | Douglas Craig | not listed |
| CHIEF RADIO ELECTRICIAN (AIR) | | | | | |
| 39272 | 29.06.51 | L/FX114539 | CHAPMAN | David Stephen | not listed |
| CHIEF ELECTRICIAN (T.I.) | | | | | |
| 39272 | 29.06.51 | P/MX766637 | BULL | Henry William | not listed |
| CHIEF ELECTRICIANS | | | | | |
| 40109 | 19.02.54 | D/MX844449 | BAISH | Kendall | HMS Consort |
| 40011 | 06.11.52 | C/MX804855 | OSBORNE | Major Percy | HMS Birmingham |
| CHIEF ELECTRICIAN (AIR) | | | | | |
| 39272 | 29.06.51 | L/FX77153 | HEADON | Clarence Frederick | not listed |
| CHIEF MECHANICIANS | | | | | |
| 39138 | 02.02.51 | D/KX88671 | EDWARDS, BEM | Headley Willis | not listed |
| 39547 | 23.05.52 | D/KX88472 | LOWTHER | Lesley Edward | not listed |
| CHIEF PETTY OFFICER TELEGRAPHISTS | | | | | |
| 39660 | 03.10.52 | C/JX138307 | CARLOW | Thomas Edward | HMS Glory |
| 39272 | 29.06.51 | P/JX113820 | DOE | Albert George | not listed |
| 39660 | 03.10.52 | C/JX135515 | HEMSLEY | Hubert Kitchener | HMS Cossack |
| 39272 | 29.06.51 | P/JX148600 | MAIRIS | Valentine Geoffrey | not listed |
| 39138 | 02.02.51 | C/JX151602 | NASH | Edward George Leonard | not listed |

| London Gazette issue No. | London Gazette date | Personal Service number | Name | Christian names | Ship's title |
|---|---|---|---|---|---|
| MENTION IN DESPATCHES (FOR GALLANT AND DISTINGUISHED SERVICES) (continued) | | | | | |
| BRONZE OAK LEAF EMBLEM | | | | | |
| CHIEF PETTY OFFICER TELEGRAPHISTS (continued) | | | | | |
| 39272 | 29.06.51 | D/JX135278 | ROGERS | William Robert Leslie | not listed |
| 39870 | 01.06.53 | D/JX133977 | TURNER | Graham Leslie | HMS Cardigan Bay |
| 39660 | 03.10.52 | D/JX153160 | WILLIAMS | Peter Allen | HMS Black Swan |
| CHIEF SHIPWRIGHT ARTIFICER | | | | | |
| 40011 | 06.11.52 | P/MX58222 | BOUSKILL | Albert Edward | HMS Birmingham |
| CHIEF ORDNANCE ARTIFICERS | | | | | |
| 39725 | 23.12.52 | P/MX47816 | BECK | Reuben Percival | HMS Ceylon |
| 40011 | 06.11.52 | P/MX51627 | ELSMORE, BEM | William John | HMS Newcastle |
| 39870 | 01.06.53 | C/MX55282 | LANE | Donald | HMS Constance |
| 39725 | 23.12.52 | C/MX47994 | WYATT | George David Albert | HMS Belfast |
| CHIEF AIRCRAFT ARTIFICERS | | | | | |
| 39854 | 19.05.53 | L/FX76931 | HAMON | Ian Ivor Basil Pearse | HMS Ocean |
| 39854 | 19.05.53 | L/FX777897 | WEBB | Frank | HMS Ocean |
| 39854 | 19.05.53 | L/FX75325 | WYNNE | David William | HMS Ocean |
| CHIEF AIR ARTIFICER (ORDNANCE) | | | | | |
| 39272 | 29.06.51 | L/FX882309 | HORNBUCKLE | Ewart Leslie | not listed |
| CHIEF AIRMAN FITTERS (E) | | | | | |
| 39682 | 28.10.52 | L/FX75012 | DUBBER | Clifford Frank | HMAS Sydney (on loan) |
| 39682 | 28.10.52 | L/FX92349 | WINSTANLEY | Arthur | HMAS Sydney (on loan) |
| CHIEF PETTY OFFICERS STORES | | | | | |
| 40109 | 19.02.54 | D/MX48786 | BEACH | Rowland Percy | HMS Sparrow |
| 39870 | 01.06.53 | P/MX58580 | BURBIDGE | Derrick James | HMS Newcastle |
| 39725 | 23.12.52 | C/MX52050 | SMITH | James Henry | HMS Belfast |
| 40109 | 19.02.54 | D/MX53487 | WILLIAMS | Henry Roy Ronald | HMS Ocean |
| PETTY OFFICER STOKER MECHANICS | | | | | |
| 39854 | 19.05.53 | P/KX82269 | HICKMAN | Leslie Ernest Elishs | HMS Whitesand Bay |
| 39725 | 23.12.52 | C/KX84595 | SMITH | John | HMS Glory |
| PETTY OFFICER TELEGRAPHISTS | | | | | |
| 39725 | 23.12.52 | P/JX160884 | DUDLEY | Reginald Lewis | HMS Whitesand Bay |
| 40109 | 19.02.54 | D/JX292653 | OAKES | Fred | not listed |
| 39272 | 29.06.51 | P/JX163506 | PAY | Maurice Eric | not listed |
| 39138 | 02.12.51 | P/JX246348 | POMEROY | Ronald George | not listed |
| 39870 | 01.06.53 | C/JX292434 | TREW | Derek Astley | HMS Birmingham |

| London Gazette issue No. | London Gazette date | Personal Service number | Name | Christian names | Ship's title |
|---|---|---|---|---|---|
| MENTION IN DESPATCHES (FOR GALLANT AND DISTINGUISHED SERVICES)(continued) | | | | | |
| BRONZE OAK LEAF EMBLEM | | | | | |
| STORES PETTY OFFICERS | | | | | |
| 39138 | 02.02.51 | D/MX63624 | HARRIS | Frank Alfred Stanley | not listed |
| 40011 | 06.11.52 | P/MX50669 | PEGLER | Herbert Alfred Worton | HMS Whitesand Bay |
| 39725 | 23.12.52 | D/MX63752 | THOMAS | William Robert | HMS Amethyst |
| 39547 | 23.05.52 | P/MX58575 | SMITH | Cyril Ewart | not listed |
| PETTY OFFICER AIR FITTERS (A) | | | | | |
| 39725 | 23.12.52 | L/FX76441 | DOWLER | Horace | HMS Glory |
| 39725 | 23.12.52 | L/FX669947 | LEITCH | James Saunders | HMS Glory |
| PETTY OFFICER AIRMAN | | | | | |
| 39138 | 02.02.51 | L/FX670436 | PRIOR | Arthur Jack | not listed |
| PETTY OFFICER COOK (S) | | | | | |
| 39854 | 19.05.53 | P/MX60633 | CURLION-COOKE | Gordon Wesley | HMS Comus |
| PETTY OFFICER STEWARD | | | | | |
| 39854 | 19.05.53 | O.1623 | KU | Hsin Kieh | HMS Cardigan Bay |
| PETTY OFFICERS | | | | | |
| 40109 | 19.02.54 | D/JX149521 | CROWE | Frederick Charles | HMS Crane |
| 40109 | 19.02.54 | P/JX581056 | ERRINGTON | Charles Bernard | HMS Newcastle |
| 39870 | 01.06.53 | P/JX141067 | HOIT | Frederick Albert | not listed |
| 40011 | 06.11.52 | C/JX667346 | McKIDDIE | Charles | HMS Glory |
| 39138 | 02.02.51 | D/JX157583 | MESSHAM | Henry | not listed |
| 39725 | 23.12.52 | C/JX292671 | TOWERS | William Hinton | HMS Belfast |
| 40109 | 19.02.54 | P/JX171715 | WILCOCK | Bernard John | HMS Morecambe Bay |
| ENGINE ROOM ARTIFICERS | | | | | |
| 39138 | 02.02.51 | C/MX63316 | COOK | Jack | not listed |
| 39854 | 19.05.53 | D/MX53099 | DIXON | Leslie Charles Alfred | HMS Consort |
| SHIPWRIGHT ARTIFICER | | | | | |
| 40011 | 06.11.52 | P/MX63550 | ABBATT | Ernest John Douglas | HMS Morecambe Bay |
| ORDNANCE ARTIFICERS | | | | | |
| 39854 | 15.05.53 | P/MX57246 | COX | James Albert | HMS Concord |
| 39854 | 15.05.53 | P/MX60664 | KIRK | Alan Frank | HMS Charity |
| RADIO ELECTRICAL ARTIFICERS | | | | | |
| 39138 | 30.01.51 | C/MX713796 | De La PLAIN | Brian Anthony | not listed |
| 39854 | 19.05.53 | P/MX766387 | JOLLY | Robert | HMS Comus |
| 39854 | 19.05.53 | L/FX594239 | LUCKEN | John Edward | HMS Ocean |

| London Gazette issue No. | London Gazette date | Personal Service number | Name | Christian names | Ship's title |
|---|---|---|---|---|---|
| MENTION IN DESPATCHES (FOR GALLANT AND DISTINGUISHED SERVICES)(continued) | | | | | |
| BRONZE OAK LEAF EMBLEM | | | | | |
| AIRCRAFT ARTIFICERS | | | | | |
| 40011 | 06.11.52 | L/FX100863 | GARRARD | Cyril Owen | HMS Unicorn |
| 39725 | 23.12.52 | L/FX87576 | MITCHELL | Leonard Matthew | HMS Glory |
| 39547 | 23.05.52 | L/FX114614 | O'BRIEN | Gerald | not listed |
| 40011 | 06.11.52 | L/FX669459 | TUFFIN | Harold Sidney | HMS Glory |
| MASTER AT ARMS | | | | | |
| 40109 | 19.02.54 | D/MX509650 | HOWELLS | Ivor George | HMS Ocean |
| YEOMEN OF SIGNALS | | | | | |
| 40011 | 06.11.52 | C/JX158811 | HOWARD | Arthur Neville | HMS Birmingham |
| 39854 | 19.05.53 | P/JX153367 | MAYERS | George Patrick | HMS Morecambe Bay |
| 39272 | 29.06.51 | P/JX152725 | MILLS, DSM | John William Harold | not listed |
| 39725 | 23.12.52 | C/JX141321 | SUTTLE | Leslie | HMS Mounts Bay |
| 39725 | 23.12.52 | P/JX245539 | TOPLEY | Albert Charles | HMS Ceylon |
| 40011 | 06.11.52 | D/JX381334 | VEAL | Jack | HMS Opossum |
| LEADING STOKER MECHANIC | | | | | |
| 39547 | 23.05.52 | P/KX164501 | RICH | Albert Lindley | not listed |
| LEADING TELEGRAPHISTS | | | | | |
| 40011 | 06.11.53 | C/JX163135 | BOWMAN | James Augustus | HMS Mounts Bay |
| 39138 | 02.02.51 | C/JX712109 | BURDALL | Derrick Walter | not listed |
| LEADING AIRMAN PILOT'S MATE | | | | | |
| 39725 | 23.12.52 | L/SFX772153 | McMICHAEL | Kenneth Arthur | HMS Glory |
| LEADING AIR MECHANIC (A) | | | | | |
| 39870 | 01.06.53 | L/FX110089 | MARK | Albert | HMS Glory |
| LEADING AIRMAN MECHANICS (O) | | | | | |
| 39725 | 23.12.52 | L/FX666625 | DAILY | Ronald | HMS Glory |
| 39854 | 19.05.53 | L/FX772332 | FOUNTAIN | Reginald Arthur | HMS Ocean |
| LEADING AIRMEN | | | | | |
| 40011 | 06.11.53 | L/FX712652 | LAW | Alexander Jones | HMS Glory |
| 39547 | 23.05.52 | L/FX838485 | RUSSELL | Alfred John | not listed |
| 39272 | 29.01.51 | L/SFX817564 | SMITH | Brian Warwick | not listed |
| LEADING STORES ASSISTANT (V) | | | | | |
| 39854 | 19.05.53 | D/SMX727994 | KING | Patrick Joseph | HMS Cardigan Bay |
| LEADING COOK (S) | | | | | |
| 39854 | 19.05.53 | D/SMX772031 | HAYNES | James | HMS St Brides Bay |
| 39272 | 29.06.51 | D/MX56537 | SMITH | Joseph Arthur | not listed |
| LEADING STEWARD | | | | | |
| 40109 | 19.02.54 | Hong Kong 0.1836 | QUAI | Ng Ah | HMS Sparrow |

| London Gazette issue No. | London Gazette date | Personal Service number | Name | Christian names | Ship's title |
|---|---|---|---|---|---|
| MENTION IN DESPATCHES (FOR GALLANT AND DISTINGUISHED SERVICES)(continued) | | | | | |
| BRONZE OAK LEAF EMBLEM | | | | | |
| LEADING SEAMEN | | | | | |
| 39660 | 03.10.52 | D/SSX795769 | COLLINGS | Kenneth Thomas | HMS St Brides Bay |
| 40011 | 06.11.52 | P/JX170181 | DUNKLEY | Ronald Lawrence | HMS Morecambe Bay |
| 39272 | 29.06.51 | C/JX804965 | FLOOK | Arthur Henry | not listed |
| 39230 | 18.05.52 | P/JX646206 | FUTCHER | Arthur John | not listed |
| 39854 | 19.05.53 | C/JX171689 | HARVEY | Thomas | HMS Mounts Bay |
| 39547 | 23.05.52 | C/JX153547 | HUTCHISON | Frederick George | not listed |
| 39660 | 03.10.52 | D/SSX840497 | ROBERTS | Stanley John | HMS St Brides Bay |
| ABLE SEAMEN | | | | | |
| 39272 | 29.06.51 | P/SSX840694 | ATKINSON | John Emery | not listed |
| 39854 | 19.05.53 | D/JX161185 | BOOTH | Thomas Albert | HMS Consort |
| 39541 | 23.05.52 | D/SSX849225 | BROWN | Sidney | not listed |
| 39725 | 23.12.52 | P/SSX838994 | BURBIDGE | Anthony Albert | HMS Concord |
| 39660 | 03.10.52 | D/SSX818522 | JARDINE | William Donald Bruce | HMS Cardigan Bay |
| 39660 | 03.10.52 | P/JX819579 | LOWE, AM | Alfred Raymond | HMS Concord |
| 39660 | 03.10.52 | D/SSX815430 | McALEENAN | Thomas Henry Alphonsus | HMS Black Swan |
| 39854 | 19.05.53 | D/SSX795784 | McDONALD | David John Kelly | HMS Cardigan Bay |
| 40109 | 19.02.54 | P/JX871695 | MASTERS | Stanley William | HMS Whitesand Bay |
| 39854 | 19.05.53 | D/SSX661848 | PHILLIPS | Donald | HMS Amethyst |
| 40011 | 06.11.53 | C/SSX841152 | RITCHIE | Ivan | HMS Mounts Bay |
| 39138 | 02.02.51 | P/JX632953 | TAYLOR | Alexander | not listed |
| 40011 | 06.11.53 | C/JX161187 | TODD | Arthur | HMS Birmingham |
| STOKER MECHANIC | | | | | |
| 40109 | 19.02.54 | C/KX893101 | JOHNSTONE | Robert McIntyre | HMS Cossack |
| TELEGRAPHIST | | | | | |
| 39547 | 23.05.52 | D/SSX843026 | APPLEBY | Terence Walter | not listed |
| ELECTRICIAN | | | | | |
| 39725 | 23.12.52 | P/MX801889 | REYNOLDS | Charles John Henry | HMS Ceylon |
| RADIO ELECTRICIANS | | | | | |
| 39547 | 23.05.52 | P/MX735168 | COCKER | Hilton Bertie | not listed |
| 39725 | 23.12.52 | C/MX844461 | POWER | John Charles | HMS Constance |
| AIRCRAFT ARTIFICER 2ND CLASS (A/E) | | | | | |
| 39272 | 29.06.51 | L/FX75046 | GATTRELL | Ronald Frederick John | not listed |
| AIRCRAFT MECHANICIAN FIRST CLASS | | | | | |
| 39854 | 19.05.53 | L/FX76154 | SAMPSON | Thomas Lewin | HMS Ocean |
| AIRCREWMAN | | | | | |
| 39138 | 02.02.51 | L/FX77297 | O'NION | Gilbert Charles | not listed |

| London Gazette issue No. | London Gazette date | Personal Service number | Name | Christian names | Ship's title |
|---|---|---|---|---|---|
| MENTION IN DESPATCH (FOR GALLANT AND DISTINGUISHED SERVICES) (continued) | | | | | |
| BRONZE OAK LEAF EMBLEM | | | | | |
| BOY FIRST CLASS | | | | | |
| 39272 | 29.06.51 | P/SSX865412 | McLEAN | Henry McGraw | not listed |
| QUEEN'S COMMENDATION (FOR COURAGE AND FORTITUDE WHILST A PRISONER OF WAR IN NORTH KOREA) | | | | | |
| BRONZE OAK LEAF EMBLEM | | | | | |
| LIEUTENANT (E) | | | | | |
| 40140 | 06.04.54 | | MATHER | Derek Graham | not listed |

| London Gazette issue No. | London Gazette date | Personal Service number | Name | Christian names | Ship's title |
|---|---|---|---|---|---|

AWARDS CONFERRED BY THE PRESIDENT OF THE UNITED STATES OF AMERICA

| London Gazette issue No. | London Gazette date | Personal Service number | Name | Christian names | Ship's title |
|---|---|---|---|---|---|
| **LEGION OF MERIT, DEGREE OF COMMANDER** | | | | | |
| VICE-ADMIRALS | | | | | |
| 39999 | 30.10.53 | | ANDREWES, KBE, CB, DSO | Sir William Gerrard | |
| 39999 | 30.10.53 | | SCOTT-MONCRIEFF, CB, CBE, DSO | Alan Kenneth | |
| **LEGION OF MERIT, DEGREE OF OFFICER** | | | | | |
| REAR-ADMIRAL | | | | | |
| 40253 | 13.08.54 | | TORLESSE, CB, DSO | Arthur David | HMS Triumph |
| COMMODORE | | | | | |
| 40253 | 13.08.54 | | ST CLARE-FORD, Bt., DSO | Sir Aubrey | HMS Belfast |
| CAPTAINS | | | | | |
| 40406 | 11.02.55 | | ADAIR, DSO, OBE | Walter Alexander | HMS Cossack |
| 40406 | 11.02.55 | | BUTLER, DSC | Adrian Rothwell Lane | HMS Amethyst |
| 40406 | 11.02.55 | | FARNOL, DSC | James Jeffrey Edward | HMS Morecambe Bay |
| 40253 | 13.08.54 | | FREWEN | John Byng | HMS Mounts Bay |
| 40253 | 13.08.54 | | JAY, DSO, DSC | Alan David Hastings | HMS Black Swan |
| 40406 | 11.02.55 | | MARSH, DSO | Richard Lovatt Haslam | HMS Crane |
| 40253 | 13.08.54 | | PODGER | Theodore Edward | HMS Kenya |
| 40253 | 13.08.54 | | THOMPSON | John Yelverton | HMS Unicorn |
| COMMANDER | | | | | |
| 40253 | 13.08.54 | | KIGGELL, DSC | Launcelot John | HMAS Sydney (on loan) |
| LIEUTENANT-COMMANDERS | | | | | |
| 40253 | 13.08.54 | | APPLEBY | John Leslie | No.808 Air Squadron |
| 40253 | 13.08.54 | | LUNBERG | Ronald Bruce | No.817 Air Squadron |
| **LEGION OF MERIT, DEGREE OF LEGIONNAIRE** | | | | | |
| CAPTAINS | | | | | |
| 40406 | 11.02.55 | | BAYLY, DSC | Patrick Uniacke | HMS Alacrity |
| 40406 | 11.02.55 | | KIMPTON, DSC | John Townshend | HMS Cockade |
| 40406 | 11.02.55 | | MILLS, DSC | Charles Piercy | HMS Concord |
| 40406 | 11.02.55 | | ROWE, DSC | George Barton | HMS Consort |
| 40253 | 13.08.54 | | UNWIN, DSC | John Harold | HMS Mounts Bay |
| SURGEON CAPTAIN | | | | | |
| 39999 | 30.10.53 | | LYNAGH, MB, Ch.B | Thomas Bernard | not listed |
| COMMANDERS | | | | | |
| 40406 | 11.02.55 | | CRAIG-WALLER, DSC | Michael Waller Beaufort | HMS Whitesand Bay |
| 40406 | 11.02.55 | | LYLE | Anthony Vacy | HMS Constance |
| 40406 | 11.02.55 | | POND | Norman Henry | HMS Ladybird |
| 40406 | 11.02.55 | | WESTERN | James Guy Trench | HMS St Brides Bay |

| London Gazette issue No. | London Gazette date | Personal Service number | Name | Christian names | Ship's title |
|---|---|---|---|---|---|
| SILVER STAR | | | | | |
| VICE-ADMIRAL | | | | | |
| 39372 | 29.06.51 | | ANDREWES, KBE, CB, DSO | Sir William Gerrard | |
| BRONZE STAR WITH COMBAT 'V' | | | | | |
| CAPTAIN | | | | | |
| 40253 | 13.08.54 | | FARNOL, DSC | James Jeffrey Edward | HMS Morecambe Bay |
| LIEUTENANT-COMMANDER | | | | | |
| 40253 | 13.08.54 | | RODWELL, DSC | Willam Edward Hunter | not listed |
| BRONZE STAR | | | | | |
| CAPTAINS | | | | | |
| 39999 | 30.10.53 | | BARBER | Hugh Shenfield | not listed |
| 39272 | 29.06.51 | | BROCK | Patrick Willet | not listed |
| 39470 | 15.02.52 | | BROWN, OBE, DSC | Walter Leslie Mather | not listed |
| 39272 | 29.06.51 | | SALTER, DSO, OBE | Jocelyn Stuart Cambridge | not listed |
| COMMANDER | | | | | |
| 40253 | 13.08.54 | | JOY | Charles Henry Francis | not listed |

# ROYAL FLEET AUXILIARY

| London Gazette issue No. | London Gazette date | Personal Service number | Name | Christian names | Ship's title |
|---|---|---|---|---|---|
| OBE: OFFICER OF THE ORDER OF THE BRITISH EMPIRE | | | | | |
| CAPTAIN | | | | | |
| 39272 | 26.06.51 | | HILL | Rowland Kelsey | Master RFA Wave Laird |
| MENTION IN DESPATCHES (FOR GALLANT AND DISTINGUISHED SERVICES) | | | | | |
| BRONZE OAK LEAF EMBLEM | | | | | |
| CAPTAINS | | | | | |
| 40011 | 06.11.52 | | COLBOURNE | Henry Fursey | Master RFA Wave Prince |
| 39854 | 19.05.53 | | HOLT | Francis Cecil | Master RFA Wave Sovereign |
| 39547 | 23.05.52 | | HUMPHREY, ODB, DSC | John Matthew | Master RFA Wave Premier |
| 39660 | 03.10.52 | | KERNICK | Stanley Charles | Master RFA Fort Rosilie |
| 39547 | 23.05.52 | | PAYNE | Eric | Master RFA Green Ranger |
| 39660 | 03.10.52 | | SHAW | Frank Arthur | Master RFA Wave Chief |
| COMMANDER | | | | | |
| 39854 | 19.05.53 | | CURTAIN, OBE, RNR | Albert Edward | Master RFA Wave Chief |
| CHIEF ENGINEERS | | | | | |
| 39547 | 23.05.52 | | MR. KENNEDY | John Edward | RFA Wave Premier |
| 39660 | 03.10.52 | | MR. PUTT | Ronald Charles | RFA Wave Chief |

# 41 INDEPENDENT COMMANDO

## ROYAL MARINES

| London Gazette issue No. | London Gazette date | Personal Service number | Name | Christian names | Remarks |
|---|---|---|---|---|---|
| 41 INDEPENDENT COMMANDO, ROYAL MARINES | | | | | |
| DSO:DISTINGUISHED SERVICE ORDER | | | | | |
| LIEUTENANT-COLONEL | | | | | |
| 39230 | 18.05.51 | not listed | DRYSDALE, MBE | Douglas Burns | |
| CAPTAIN | | | | | |
| 39660 | 03.10.52 | not listed | SHULDHAM | Edward Thomas Gahan | |
| DSC:DISTINGUISHED SERVICE CROSS | | | | | |
| LIEUTENANT | | | | | |
| 39660 | 03.10.52 | not listed | WALTER | John Roche Hensleigh | |
| MC:MILITARY CROSS | | | | | |
| MAJOR | | | | | |
| 39320 | 18.05.51 | not listed | ALDRIDGE, MBE | Dennis Leolin Samuel St Maur | |
| CAPTAINS | | | | | |
| 39320 | 18.05.51 | not listed | MARSH | Leslie George | |
| 39320 | 18.05.51 | not listed | OVENS | Patrick John | |
| DSM:DISTINGUISHED SERVICE MEDAL | | | | | |
| QUARTERMASTER-SERGEANT | | | | | |
| 39660 | 03.10.52 | CH/X2480 | DODDS, MM | Reginald Roy | listed as prisoner of war |
| LEADING SICK BERTH ATTENDANT | | | | | |
| 39660 | 03.10.52 | D/MX764874 | CRITCHLEY | Evan | |
| MM:MILITARY MEDAL | | | | | |
| COLOUR SERGEANT | | | | | |
| 39230 | 18.05.51 | PLY/X920 | BAINES, BEM | James | |
| SERGEANT | | | | | |
| 39230 | 18.05.51 | PO/X2579 | JAMES | Reginald William David | |
| CORPORALS | | | | | |
| 39230 | 18.05.51 | PLY/X4655 | CRUSE | Ernest | |
| 39230 | 18.05.51 | CH/X4045 | LANGTON | Henry | |
| 39230 | 18.05.51 | CH/X4387 | MAINDONALD | Gersham | |
| MARINES | | | | | |
| 39230 | 18.05.51 | CH/X114851 | BRAMBLE | George | |
| 39230 | 18.05.51 | PO/X6207 | HARPER | Arthur Alexander Henry | |
| 39230 | 18.05.51 | RM8381 | HINE | Malcolm | |
| 39230 | 18.05.51 | RM8351 | TWIGG | Richard | |

| London Gazette issue No. | London Gazette date | Personal Service number | Name | Christian names | Remarks |
|---|---|---|---|---|---|
| MENTION IN DESPATCHES (FOR GALLANT AND DISTINGUISHED SERVICES) | | | | | |
| BRONZE OAK LEAF EMBLEM | | | | | |
| LIEUTENANTS | | | | | |
| 39230 | 10.05.51 | not listed | ROBERST | Gerald Frederick Dawson | |
| 39230 | 10.05.51 | not listed | THOMAS | Peter Roome | |
| SERGEANTS | | | | | |
| 39230 | 10.05.51 | PO/X4003 | BARNES | Charles Ernest | |
| 39230 | 10.05.51 | RM7429 | MACKAY | William Muir | |
| CORPORAL | | | | | |
| 39230 | 10.05.51 | PLY/X5202 | TORR | Derek Austin | |
| MARINES | | | | | |
| 39660 | 03.10.52 | CH/X114854 | BRAMBLE, MM | George | |
| 39230 | 18.05.51 | RM8747 | ROBERTS | Frederick | |
| 39230 | 18.05.51 | PL/X5553 | VARDY | Stanley | |
| QUEEN'S COMMENDATION | | | | | |
| (FOR COURAGE AND FORTITUDE WHILST A PRISONER OF WAR IN NORTH KOREA) | | | | | |
| BRONZE OAK LEAF EMBLEM | | | | | |
| QUARTERMASTER SERGEANT | | | | | |
| 40140 | 06.04.54 | CH/X1206 | DAY | James | |

| London Gazette issue No. | London Gazette date | Personal Service number | Name | Christian names | Remarks |
|---|---|---|---|---|---|
| AWARDS CONFERRED BY THE PRESIDENT OF THE UNITED STATES OF AMERICA | | | | | |
| LEGION OF MERIT, DEGREE OF OFFICER | | | | | |
| MAJOR | | | | | |
| 40253 | 13.08.54 | not listed | GRANT | Ferris Nelson | |
| SILVER STAR | | | | | |
| LIEUTENANT-COLONELS | | | | | |
| 40253 | 13.08.54 | not listed | ALDRIDGE, MBE, MC | Dennis Leolin Samuel St Maur | |
| 39272 | 29.06.51 | not listed | DRYSDALE, DSO, MBE | Douglas Burns | Two Silver Stars shown in Special Publication No. 8 of Royal Marines Historical Society: "41 Independent Commando RM, Korea 1950-52," but only one entry traced in London Gazette. |
| BRONZE STAR WITH COMBAT 'V' INSIGNIA | | | | | |
| CAPTAINS | | | | | |
| 40253 | 13.08.54 | not listed | POUNDS | Edgar George Derrick | |
| 40253 | 13.08.54 | not listed | THOMAS | Peter Roome | |
| LIEUTENANT | | | | | |
| 39328 | 13.08.54 | not listed | ROBERTS | Gerald Fredrick Dawson | |
| WARRANT OFFICER CLASS I (RSM) | | | | | |
| | | PLY/X920 | BAINES, MM, BEM | James | This entry taken from Special Publication No.8 of Royal Marines Historical Society: "41 Independent Commando RM, Korea 1950-52". |
| QUARTERMASTER SERGEANT | | | | | |
| 40253 | 13.08.54 | CH/X1698 | HAMBLETON | Charles Frederick | |
| SERGEANTS | | | | | |
| 40253 | 13.08.54 | CH/X4248 | MAILL | Joseph Jermiah | |
| 40253 | 13.08.54 | PO/X5235 | MOON | Sidney James | |
| 40253 | 13.08.54 | PO/X4066 | WHITING | Jack | |
| CORPORAL | | | | | |
| 40253 | 13.08.54 | PLY/X5068 | BAKER | Roy | |
| MARINE | | | | | |
| 40253 | 13.08.54 | RM9043 | LANDSDOWN | Arthur John | |

ROYAL MARINES

| London Gazette issue No. | London Gazette date | Personal Service number | Name | Christian names | Ship's title |
|---|---|---|---|---|---|
| ROYAL MARINES | | | | | |
| | | | | | |
| DSC: DISTINGUISHED SERVICE CROSS | | | | | |
| CAPTAINS | | | | | |
| 39660 | 03.10.52 | not listed | BURGE | Humphrey Edward Kelsey | HMS Ceylon |
| 39547 | 20.05.52 | not listed | DILLON | John Desmond | not listed |
| | | | | | |
| DSM: DISTINGUISHED SERVICE MEDAL | | | | | |
| CORPORALS | | | | | |
| 39854 | 19.05.53 | CH/X4062 | TONG | Gordon Leslie | HMS Belfast |
| 39660 | 03.10.52 | PO/X3521 | WEBSTER | William James | HMS Ceylon |
| | | | | | |
| BEM: BRITISH EMPIRE MEDAL | | | | | |
| SERGEANT | | | | | |
| 39138 | 30.01.51 | PO/X2351 | WATT | John | not listed |
| | | | | | |
| MENTION IN DESPATCHES (FOR GALLANT AND DISTINGUISHED SERVICES) | | | | | |
| BRONZE OAK LEAF EMBLEM | | | | | |
| BANDMASTERS | | | | | |
| 39547 | 23.05.52 | RMB/X189 | BUCKINGHAM | Ernest William | not listed |
| 39725 | 23.12.52 | RMB/X360 | SPENCER | Walter James | HMS Glory |
| COLOUR-SERGEANTS | | | | | |
| 39272 | 29.06.51 | PLY/X5125 | HENSEY | Alfred Harold | not listed |
| 40011 | 06.11.53 | PLY/X1555 | SULLY | Albert Reginald | HMS Newcastle |
| SERGEANTS | | | | | |
| 39272 | 29.06.51 | PO/X1181 | JURY | Reginald Leonard | not listed |
| 39138 | 02.02.51 | PLY/X2634 | THELWELL | William Robert James | not listed |
| CORPORAL | | | | | |
| 39725 | 23.12.52 | RM8416 | HARKER | Frederick Noel | HMS Ceylon |
| MARINE | | | | | |
| 39725 | 23.12.52 | RM7345 | MANSELL | David John Stanley | HMS Ceylon |

## AWARDS CONFERRED BY THE PRESIDENT OF THE UNITED STATES OF AMERICA

| London Gazette issue No. | London Gazette date | Personal Service number | Name | Christian names | Ship's title |
|---|---|---|---|---|---|
| SILVER STAR | | | | | |
| CAPTAIN | | | | | |
| 39999 | 27.10.53 | not listed | BURGE, DSC | Humphrey Edward Kelsey | HMS Ceylon |

# THE MILITARY

| London Gazette issue No. | London Gazette date | Personal Service number | Name | Christian names | Remarks |
|---|---|---|---|---|---|
| HEADQUARTERS (STAFF) | | | | | |
| KCB: KNIGHT COMMANDER OF THE ORDER OF THE BRITISH EMPIRE | | | | | |
| MAJOR-GENERAL | | | | | |
| 39555 | 10.10.52 | 36316 | CASSELS, CB, CBE, DSO | Archibald James Halkett | Commander 1st Commonwealth Division |
| CBE: COMMANDER OF THE ORDER OF THE BRITISH EMPIRE | | | | | |
| BRIGADIERS | | | | | |
| 39084 | 05.12.50 | 34467 | COAD, DSO | Basil Aubrey | Commander 27 Comwel Infantry Brigade |
| 40036 | 04.12.53 | 34436 | GREGSON, DSO, MC | Guy Patrick | late Royal Regiment of Artillery |
| 39666 | 10.10.52 | 31590 | PIKE, DSO | William George Huddleston | late Royal Regiment of Artillery |
| 39328 | 07.09.51 | 31598 | ROWLANDSON, OBE | John Claude | late Royal Regiment of Artillery |
| COLONEL | | | | | |
| 39870 | 26.05.53 | 36678 | HILL | Frank Mourtray | late Royal Corps of Engineers |
| 3RD BAR TO DSO: DISTINGUISHED SERVICE ORDER | | | | | |
| BRIGADIER | | | | | |
| 40036 | 04.12.53 | 44766 | KENDREW, CBE, DSO | Douglas Anthony | Commander 29 Brit Inf Brigade Group |
| 2ND BAR TO DSO: DISTINGUISHED SERVICE ORDER | | | | | |
| MAJOR-GENERAL | | | | | |
| 40036 | 04.12.53 | 33582 | ALSTON-ROBERTS-WEST, CB, DSO | Michael Montgomerie | Commander 1st Commonwealth Division |
| DSO: DISTINGUISHED SERVICE ORDER | | | | | |
| BRIGADIER | | | | | |
| 39328 | 07.09.51 | 34236 | BRODIE, CBE | Thomas | Commander 29 Brit Inf Brigade Group |
| DSO: DISTINGUISHED SERVICE ORDER (FOR FLYING OPERATIONS IN KOREA) | | | | | |
| MAJOR | | | | | |
| 39833 | 24.04.53 | 105892 | HAILES | John Martin Hunter | Royal Regiment of Artillery |
| OBE: OFFICER OF THE ORDER OF THE BRITISH EMPIRE | | | | | |
| COLONEL | | | | | |
| 39666 | 10.10.52 | 38524 | MAXWELL, DSO | John Lockhart | Scots Guards: HQ 1 Comwel Division |
| LIEUTENANT-COLONELS | | | | | |
| 39870 | 26.05.53 | 53176 | DIMSDALE, MC | Henry Cockfield Luke | Welsh Guards: HQ 1 Comwel Division |
| 39870 | 26.05.53 | 51346 | McCONNELL | David William | King's Own Scottish Borderers |
| 40036 | 04.12.53 | 52726 | TAITE | Richard Hamish | King's Own Royal Regiment |

| London Gazette issue No. | London Gazette date | Personal Service number | Name | Christian names | Remarks |
|---|---|---|---|---|---|
| OBE: OFFICER OF THE ORDER OF THE BRITISH EMPIRE (continued) | | | | | |
| MAJORS | | | | | |
| 39212 | 24.04.51 | 69090 | BEAUMONT-NESBITT, MBE | David John Frederick | Grenadier Guards |
| 39528 | 29.04.52 | 63586 | TREVOR, DSO | Kenneth Roland Swetenam | Cheshire Regiment |
| MBE: MEMBER OF THE ORDER OF THE BRITISH EMPIRE | | | | | |
| MAJORS | | | | | |
| 39870 | 26.05.53 | 279935 | DUNCAN | James Sidney | RASC - Expeditionary Forces Institute |
| 39831 | 21.04.53 | 74662 | EARLE | Albert Vivian | South Lancashire Regiment |
| 39831 | 21.04.53 | 159355 | FARRELL | Donald Patrick | Royal Inniskilling Fusiliers |
| 40036 | 04.12.53 | 62647 | KING-CLARK, MC | Robert | Manchester Regiment |
| 40036 | 04.12.53 | 66173 | MACRAE | Robert Andrew Alexander Scarth | Seaforth Highlanders |
| 39831 | 21.04.53 | 64576 | RAWLINGS | Anthony Clive | The Buffs |
| 39831 | 21.04.53 | 263746 | WEST, MC | Thomas John | Royal Irish Fusiliers |
| 39666 | 10.10.52 | 62591 | YOUNG | Guy Ronald | Devonshire Regiment |
| CAPTAINS | | | | | |
| 39870 | 01.06.53 | 366651 | BIRNIE | Albert Wyness | RASC - Expeditionary Forces Institute |
| 40036 | 04.12.53 | 273889 | FAREHAM | John Henry | Sherwood Foresters |
| WARRANT OFFICER CLASS II | | | | | |
| 39528 | 29.04.52 | 5670430 | GRAY | Harold Charles James | Somerset Light Infantry |
| MC: MILITARY CROSS | | | | | |
| MAJOR | | | | | |
| 40036 | 04.12.53 | 75264 | TAYLOR | John Aveline Waters | King's Own Royal Regiment |
| DFC: DISTINGUISHED FLYING CROSS | | | | | |
| MAJOR | | | | | |
| 39661 | 03.10.52 | 271491 | GOWER | Ronald Norton Leveson | Royal Regiment of Artillery |

| London Gazette issue No. | London Gazette date | Personal Service number | Name | Christian names | Remarks |
|---|---|---|---|---|---|
| DFC: DISTINGUISHED FLYING CROSS (continued) | | | | | |
| CAPTAINS | | | | | |
| 39505 | 01.04.52 | 277301 | ADDINGTON | Leslie Richard Bagnall | Royal Regiment of Artillery 1903 Air Observation Post Flight |
| 40106 | 19.02.54 | 387271 | BROWNE | Donald John | Royal Regiment of Artillery |
| 39320 | 22.05.51 | 210702 | COX | Frederick Augustine | Royal Regiment of Artillery |
| 39833 | 24.04.53 | 373619 | CRAWSHAW | John Arthur | Royal Regiment of Artillery |
| 39661 | 03.10.52 | 299293 | DOWNWARD | Peter Aldcroft | South Lancashire Regiment 1913 Light Liaison Flight AOP RAF |
| 39230 | 22.05.51 | 219199 | HALL | Denis Arthur | Royal Regiment of Artillery |
| 39661 | 03.10.52 | 364933 | JARVIS | Derek Bernard William | Royal Regiment of Artillery |
| 39833 | 24.04.53 | 398878 | JOYCE | Gerald William Cliff | Royal Regiment of Artillery |
| 39870 | 01.06.53 | 373740 | NICHOLLS | William Terrence Ashley | Royal Regiment of Artillery |
| 40009 | 06.11.53 | 369841 | PERKINS | Kenneth | Royal Regiment of Artillery |
| 39661 | 03.10.52 | 335387 | STEWART-COX | Arthur George Ernest | Royal Regiment of Artillery |
| 40009 | 06.11.53 | 312975 | WILSON | Peter Fitzmaurice | The Buffs |
| LIEUTENANT | | | | | |
| 40106 | 19.02.54 | 373609 | HOARE | John Edward Teague | Royal Regiment of Artillery |
| MM: MILITARY MEDAL | | | | | |
| CORPORAL | | | | | |
| 40036 | 04.12.53 | 22549787 | GARRETT | Michael Montague | Royal Sussex Regiment |
| LANCE-CORPORAL | | | | | |
| 39831 | 01.04.53 | 22542579 | HANNA | William Henry Wilson | Royal Irish Fusiliers |
| PRIVATE | | | | | |
| 39589 | 04.07.52 | 22045820 | ELLAWAY | Thomas Patrick | South Wales Borderers |
| DFM: DISTINGUISHED FLYING MEDAL | | | | | |
| STAFF-SERGEANT | | | | | |
| 40010 | 06.11.53 | 1151829 | ROLLEY | John Charles | The Glider Pilot Regiment 1913 Light Liaison Flight AOP RAF |
| SERGEANT | | | | | |
| 39505 | 01.04.52 | 14189106 | HUTCHINGS | James William | King's Royal Rifle Corps |
| BEM: BRITISH EMPIRE MEDAL | | | | | |
| WARRANT OFFICER CLASS II | | | | | |
| 40036 | 08.12.53 | 22538040 | MARTIN | Henry Albert | RASC - Expeditionary Forces Institute |
| COLOUR-SERGEANT | | | | | |
| 40036 | 08.12.53 | 3769322 | GEE | Robert Ernest | Bedfordshire and Hertfordshire Regiment |

| London Gazette issue No. | London Gazette date | Personal Service number | Name | Christian names | Remarks |
|---|---|---|---|---|---|
| BEM: BRITISH EMPIRE MEDAL (continued) | | | | | |
| SERGEANTS | | | | | |
| 39666 | 10.10.52 | 2756430 | THOMPSON | George William | Corps of Royal Military Police |
| 40036 | 08.12.53 | 22197930 | WILLIAMS | Kenneth Leslie | Corps of Royal Military Police |
| MENTION IN DESPATCHES (FOR GALLANT AND DISTINGUISHED FLYING SERVICES) | | | | | |
| BRONZE OAK LEAF EMBLEM | | | | | |
| CAPTAINS | | | | | |
| 39661 | 03.10.52 | 299655 | BROWN | A.T.C. | Gordon Highlanders |
| 40106 | 19.02.54 | 385320 | STEPTO | A.S. | Royal Regiment of Artillery |
| 39505 | 01.04.52 | 335387 | STEWART-COX | Arthur George Ernest | Royal Regiment of Artillery |
| STAFF-SERGEANT | | | | | |
| 39661 | 03.10.52 | 5127631 | HALL | R.E. | The Glider Pilot Regiment |
| SERGEANTS | | | | | |
| 40106 | 19.02.54 | 22541403 | BRENNAN | A.J. | Royal Tank Regiment |
| 39661 | 03.10.52 | 14189106 | HUTCHINGS, DFM | James William | King's Royal Rifle Corps |
| 39833 | 24.04.53 | 19030262 | KILLELEA | J.P. | Royal Inniskilling Fusiliers |
| LANCE-BOMBARDIER | | | | | |
| 39505 | 10.04.52 | 19034837 | NASH | C.K.G. | Royal Regiment of Artillery |
| MENTION IN DESPATCHES (FOR GALLANT AND DISTINGUISHED SERVICES) | | | | | |
| BRONZE OAK LEAF EMBLEM | | | | | |
| MAJOR-GENERAL | | | | | |
| 40036 | 04.12.53 | 33582 | ALSTON-ROBERTS-WEST, CB, DSO | Michael Montgomerie | Commander 1st Commonwealth Division Staff Corps: late infantry |
| COLONELS | | | | | |
| 39666 | 10.10.52 | 36781 | ANDERTON, OBE, MB | Geoffrey | Staff Corps: late Royal Army Medical Corps |
| 39831 | 24.04.53 | 39171 | HENDERSON, DSO | P.R. | Staff Corps: late Royal Regiment of Artillery |
| 39666 | 10.10.52 | 36717 | MYERS, CBE, DSO | E.C.W. | Staff Corps: late Royal Corps of Engineers |
| 40036 | 04.12.53 | 47683 | SLEEMAN, OBE | John Cuthbert de Fontenne | Staff Corps: late RAC - Royal Tank Regiment |
| LIEUTENANT-COLONEL | | | | | |
| 39666 | 10.10.52 | 62638 | VICKERS, OBE | A.W.N.L. | King's Own Yorkshire Light Infantry |
| MAJORS | | | | | |
| 39528 | 29.04.52 | 270701 | BENNETT, MBE | C. | Royal Armoured Corps attached the Corps of Royal Military Police |
| 39531 | 02.05.52 | 355863 | EVANS, MM | A.W. | Somerset Light Infantry |

| London Gazette issue No. | London Gazette date | Personal Service number | Name | Christian names | Remarks |
|---|---|---|---|---|---|
| MENTION IN DESPATCHES (FOR GALLANT AND DISTINGUISHED SERVICES) (continued) | | | | | |
| BRONZE OAK LEAF EMBLEM | | | | | |
| MAJORS (continued) | | | | | |
| 39831 | 24.04.53 | 153960 | LAWSON | P.M. | Royal Sussex Regiment |
| 39666 | 10.10.52 | 63722 | THOMAS, OBE | R.C.W. | Queen's Own Royal West Kent Regiment |
| 39328 | 07.09.51 | 63586 | TREVOR, DSO | Kenneth Roland Swetenam | Cheshire Regiment |
| CAPTAINS | | | | | |
| 39831 | 24.04.53 | 269423 | BINGLEY | R.A.J. | South Staffordshire Regiment |
| 39831 | 24.04.53 | 415121 | CASEY | J.D. | Royal Irish Fusiliers |
| 39938 | 14.08.53 | 265913 | EVANS, MC | P.S. | York and Lancaster Regiment |
| 40036 | 04.12.53 | 362049 | HAYES | P.S. | Oxfordshire and Buckinghamshire Light Infantry |
| 39831 | 24.04.53 | 262925 | LLOYD-JONES, MC | I.W. | Royal Welch Fusiliers |
| 39938 | 14.08.53 | 359890 | O'KANE | P.J. | Royal Berkshire Regiment |
| LIEUTENANTS | | | | | |
| 40036 | 04.12.53 | 373024 | ALLMAN | D.G. | Royal Hampshire Regiment |
| 39272 | 29.06.51 | 364199 | WOLFE | G.D.H. | 9th Queen's Royal Lancers |
| 2ND LIEUTENANTS | | | | | |
| 39831 | 24.04.53 | 420260 | DOIG | J.R.K. | Seaforth Highlanders listed: killed in action |
| 40036 | 04.12.53 | 422799 | HALL | S.H. | King's Own Yorkshire Light Infantry |
| 39831 | 24.04.53 | 420642 | McGUIGAN | J.G.M. | The Cameronians listed: killed in action |
| WARRANT OFFICER CLASS II | | | | | |
| 40036 | 04.12.53 | 14002749 | RULE | B.R. | East Yorkshire Regiment |
| SERGEANT | | | | | |
| 39831 | 24.04.53 | 22267596 | WESLEY | G. | East Yorkshire Regiment |
| CORPORALS | | | | | |
| 39328 | 07.09.51 | 5333610 | BALL | J.H. | Worcestershire Regiment |
| 39528 | 29.04.52 | 6980308 | BRUCE | D. | Royal Inniskilling Fusiliers |
| 40036 | 04.12.53 | 22581419 | CHEESMAN | F.E. | Queen's Royal Regiment |

| London Gazette issue No. | London Gazette date | Personal Service number | Name | Christian name | Remarks |
|---|---|---|---|---|---|
| | | | **AWARDS CONFERRED BY THE PRESIDENT OF THE UNITED STATES OF AMERICA** | | |
| **LEGION OF MERIT, DEGREE OF COMMANDER** | | | | | |
| **MAJOR-GENERALS** | | | | | |
| 40249 | 06.08.54 | 33582 | ALSTON-ROBERTS-WEST, CB, DSO | Michael Montgomerie | Commander 1st Commonwealth Division Staff Corps: late infantry |
| 39646 | 16.09.52 | 36316 | CASSELS, CB, CBE, DSO | Archibald James Halkett | Commander 1st Commonwealth Division Staff Corps: late infantry |
| 40249 | 06.08.54 | 18052 | SHOOSMITH, CB, DSO, OBE | Stephen Newton | Staff Corps: late Royal Regiment of Artillery |
| **COLONEL** | | | | | |
| 40081 | 22.01.54 | 36781 | ANDERTON, OBE, MB | Geoffrey | Staff Corps: late Royal Army Medical Corps. This Award substitutes Legion of Merit, Degree Legionnaire promulgated in London Gazette No.39999 dated 27.10.53. |
| **LEGION OF MERIT, DEGREE OF OFFICER** | | | | | |
| **MAJOR-GENERALS** | | | | | |
| 39999 | 27.10.53 | 34236 | * BRODIE, CB | Thomas | Commander 29 Indep Inf Brigade Group Staff Corps: late infantry. |
| 39999 | 27.10.53 | 34667 | * COAD, CB, CBE, DSO | Basil Aubrey | Commander 27th Commonwealth Infantry Brigade. Staff Corps * These Awards substitute the USA Silver Star promulgated in London Gazette No.39254 dated 08.06.51. |
| **BRIGADIERS** | | | | | |
| 39999 | 27.10.53 | 30523 | MacDONALD, DSO, OBE | John Frederick Matheson | Commander 28th Commonwealth Infantry Brigade. Staff Corps |
| 39999 | 27.10.53 | 31590 | PIKE, CB, DSO | William Gregory Huddleston | Staff Corps: late Royal Regiment of Artillery |
| 39999 | 27.10.53 | 33748 | RICKETTS, CBE, DSO | Abdy Henry Gough | Commander 29 Indep Inf Brigade Group Staff Corps: late infantry |
| **COLONEL** | | | | | |
| 40249 | 06.08.54 | 47683 | SLEEMAN, OBE | John Cuthbert de Fontenne | Staff Corps: late Royal Armoured Corps Royal Tank Regiment |
| **LEGION OF MERIT, DEGREE OF LEGIONNAIRE** | | | | | |
| **COLONELS** | | | | | |
| 39999 | 27.10.53 | 39171 | HENDERSON | Peter Reynolds | Staff Corps: late Royal Regiment of Artillery |

| London Gazette issue No. | London Gazette date | Personal Service number | | Name | Christian names | Remarks |
|---|---|---|---|---|---|---|
| LEGION OF MERIT, DEGREE OF LEGIONNAIRE (continued) | | | | | | |
| COLONELS (continued) | | | | | | |
| 39999 | 27.10.53 | 36718 | | MARTIN, DSO | Frederick Lawrence | Staff Corps: late infantry |
| 39999 | 27.10.53 | 36717 | | MYERS, CBE, DSO | Edmund Charles Wolf | Staff Corps: late Corps of Royal Engineers |
| SILVER STAR | | | | | | |
| MAJOR-GENERALS | | | | | | |
| 40249 | 06.08.54 | 34236 | * | BRODIE, CB, CBE, DSO | Thomas | Commander 29 Indep Inf Brigade Group Staff Corps: late infantry |
| 40249 | 06.08.54 | 34667 | * | COAD, CB, CBE, DSO | Basil Aubrey | Commander 27th Commonwealth Infantry Brigade. Staff Corps: late infantry<br>* The Silver Star awards originally conferred and promulgated in the London Gazette No.39254 dated 08.06.51 were subsequently substituted by the higher award of Legion of Merit, degree of Officer, promulgated in London Gazette No. 39999 dated 27.10.53. The Silver Star listed here, in both instances, are unrelated in antecedance to the original Silver Star awards, which were substituted. Editor. |
| BRONZE STAR | | | | | | |
| LIEUTENANT-COLONEL | | | | | | |
| 39462 | 08.02.52 | 69090 | | BEAUMONT-NESBITT, OBE | David John Frederick | Staff Corps: Grenadier Guards |
| MAJORS | | | | | | |
| 39462 | 08.02.52 | 355863 | | EVANS, MM | Albert Walker | Somerset Light Infantry |
| 39999 | 27.10.53 | 63722 | | THOMAS, OBE | Robert Cyril Wolferstan | Queen's Own Royal West Kent Regiment |
| 39999 | 27.10.53 | 62638 | | VICKERS, OBE | Arthur William Neville Langston | King's Own Yorkshire Light Infantry |
| CAPTAINS | | | | | | |
| 39999 | 27.10.53 | 370061 | | BALFOUR-SCOTT | David | Royal Scots Fusiliers |
| 39999 | 27.10.53 | 330635 | | BENTLEY | Ralph Donald Carvalho | The Life Guards |

| London Gazette issue No. | London Gazette date | Personal Service number | Name | Christian names | Remarks |
|---|---|---|---|---|---|
| U.S.A. DISTINGUISHED FLYING CROSS | | | | | |
| LIEUTENANTS | | | | | |
| 40249 | 06.08.54 | 414804 | BERRY | Peter Stanton | RAC: Royal Tank Regiment |
| 40249 | 06.08.54 | 411934 | FANSHAWE | Peter Douglas | The Queen's Bays |
| 39662 | 08.02.52 | 407957 | MARTIN | David William | 8th King's Royal Irish Hussars |
| 40249 | 06.08.54 | 385974 | WILSON | Kenneth | The Royal Scots |
| AIR MEDAL AND OAK LEAF CLUSTER | | | | | |
| CAPTAIN | | | | | |
| 39999 | 27.10.53 | 398666 | BURGESS, MM | Desmond Horace Reginald | The Black Watch |
| AIR MEDAL | | | | | |
| CAPTAINS | | | | | |
| 39999 | 27.10.53 | 314464 | BEGBIE | Robert Martin | Royal Regiment of Artillery |
| 39999 | 27.10.53 | 364933 | JARVIS, DFC | Derek Bernard William | Royal Regiment of Artillery |
| LIEUTENANTS | | | | | |
| 39999 | 27.10.53 | 390280 | BRINSON | William Lindsay Neilson | 3rd The King's Own Hussars |
| 39999 | 27.10.53 | 397825 | BROWNING | Jeffrey Edward Power | 8th King's Royal Irish Hussars |
| 39999 | 27.10.53 | 407957 | MARTIN | David William | 8th King's Royal Irish Hussars |
| 39999 | 27.10.53 | 354925 | SKOYLES | Eric Leslie | Royal Northumberland Fusiliers |
| 39999 | 27.10.53 | 373721 | TOLPUTT | Wilfred Peter Riordan | Royal Regiment of Artillery |
| SERGEANT | | | | | |
| 39999 | 27.10.53 | 14185489 | JERMY | Harold | The Glider Pilot Regiment |

| London Gazette issue No. | London Gazette date | Personal Service number | Name | Christian names | Remarks |
|---|---|---|---|---|---|

## THE ROYAL ARMOURED CORPS

### 5TH ROYAL INNISKILLING DRAGOON GUARDS

5th Dragoon Guards: Green Horse
6th Dragoons: The Skins
Date raised: Cuirassiers 1685
Inniskilling Dragoons 1689

MBE: MEMBER OF THE ORDER OF THE BRITISH EMPIRE

MAJORS

| London Gazette issue No. | London Gazette date | Personal Service number | Name | Christian names | Remarks |
|---|---|---|---|---|---|
| 39831 | 21.04.53 | 90618 | BLUNDELL-BROWN | Geoffrey Tom | |
| 39831 | 21.04.53 | 88796 | WALKER, MC | Harry Charles | |
| WARRANT OFFICER CLASS II | | | | | |
| 39831 | 21.04.53 | 7885963 | GREEN | Leonard William | |

MC: MILITARY CROSS

CAPTAIN

| London Gazette issue No. | London Gazette date | Personal Service number | Name | Christian names | Remarks |
|---|---|---|---|---|---|
| 39282 | 10.07.51 | 262386 | MURRAY | Gavin Stuart | |
| LIEUTENANTS | | | | | |
| 39749 | 09.01.53 | 407735 | ANSTICE | Michael John Christian | |
| 39831 | 21.04.53 | 40083 | TAYLOR | Charles Eyre | |

MM: MILITARY MEDAL

SERGEANT

| London Gazette issue No. | London Gazette date | Personal Service number | Name | Christian names | Remarks |
|---|---|---|---|---|---|
| 39831 | 21.04.53 | 19031442 | IRVING | John Christopher | |

BEM: BRITISH EMPIRE MEDAL

WARRANT OFFICER CLASS II

| London Gazette issue No. | London Gazette date | Personal Service number | Name | Christian names | Remarks |
|---|---|---|---|---|---|
| 39831 | 21.04.53 | 409236 | GILLIAND | William | |

MENTION IN DESPATCHES (FOR GALLANT AND DISTINGUISHED SERVICES)

BRONZE OAK LEAF EMBLEM

LIEUTENANT-COLONEL

| London Gazette issue No. | London Gazette date | Personal Service number | Name | Christian names | Remarks |
|---|---|---|---|---|---|
| 39831 | 24.04.53 | 63557 | VIGORS, DSO | R de C | |
| CAPTAIN | | | | | |
| 39666 | 10.10.52 | 320930 | WATHEN | G.L. | Royal Armoured Corps attached |
| LIEUTENANT-QUARTERMASTER | | | | | |
| 39666 | 10.10.52 | 418970 | BIRCHALL | F. | Royal Armoured Corps attached |
| WARRANT OFFICER CLASS II | | | | | |
| 39831 | 24.04.53 | 406301 | CLAYTON | J. | |
| 39831 | 24.04.53 | 807638 | GOWER | C.F. | |

| London Gazette issue No. | London Gazette date | Personal Service number | Name | Christian names | Remarks |
|---|---|---|---|---|---|
| MENTION IN DESPATCHES (FOR GALLANT AND DISTINGUISHED SERVICES)(continued) | | | | | |
| BRONZE OAK LEAF EMBLEM | | | | | |
| SERGEANTS | | | | | |
| 39666 | 10.10.52 | 19039480 | COWLING | F. | |
| 39831 | 24.04.53 | 22022059 | HUBBARD | H.A. | |
| 39666 | 10.10.52 | 14310634 | MUSK | P.A. | |
| TROOPER | | | | | |
| 39831 | 24.04.53 | 22439436 | HEDGES | G.T. | The Queen's Bays attached |

## AWARDS CONFERRED BY THE PRESIDENT OF THE UNITED STATES OF AMERICA

| London Gazette issue No. | London Gazette date | Personal Service number | Name | Christian names | Remarks |
|---|---|---|---|---|---|
| BRONZE STAR WITH COMBAT 'V' INSIGNIA | | | | | |
| SERGEANT | | | | | |
| 39999 | 27.10.53 | 19034478 | WRIGHT | Frank | |
| BRONZE STAR | | | | | |
| CAPTAIN | | | | | |
| 39999 | 27.10.53 | 326841 | CUPPER | John Colin | |

| London Gazette issue No. | London Gazette date | Personal Service number | Name | Christian names | Remarks |
|---|---|---|---|---|---|
| 8TH KING'S ROYAL IRISH HUSSARS | | | | | 8th Dragoons raised 1693<br>The Crossed Belts raised 1769 |
| DSO: DISTINGUISHED SERVICE ORDER | | | | | |
| MAJORS | | | | | |
| 39398 | 30.11.51 | 74286 | BUTLER, MC | Walter George Ormonde | |
| 39528 | 29.03.52 | 50859 | De CLERMONT | Patrick Howard Voltelin | |
| 39252 | 01.06.51 | 66068 | HUTH, MC | Percival Henry | |
| OBE: OFFICER OF THE ORDER OF THE BRITISH EMPIRE | | | | | |
| LIEUTENANT-COLONEL | | | | | |
| 39528 | 29.04.52 | 53672 | LOWTHER, Bt. | Sir William Guy | |
| MAJOR | | | | | |
| 39528 | 29.04.52 | 62563 | NELSON | William Vernon Hope | |
| MBE: MEMBER OF THE ORDER OF THE BRITISH EMPIRE | | | | | |
| LIEUTENANTS | | | | | |
| 39406 | 11.02.51 | 397198 | BUTLER | James Charles | |
| 39528 | 29.04.52 | 393242 | PAUL | Charles Edward | 14th/20th King's Hussars: 14th Dragoons; attached |
| 39528 | 29.04.52 | 403708 | TROUGHTON | Christopher Douglas Bryan | |
| WARRANT OFFICERS CLASS II | | | | | |
| 39528 | 29.04.52 | 552302 | DAY | Alfred Newall | |
| 39528 | 29.04.52 | 2928621 | FRIEL | Daniel | |
| MC: MILITARY CROSS | | | | | |
| CAPTAINS | | | | | |
| 39282 | 10.07.51 | 237398 | ORMROD | Peter Charles | |
| 39406 | 11.12.51 | 334075 | PIPER | Richard Warren | |
| 39528 | 29.04.52 | 236064 | WINN | The Hon. Rowland Denys Guy | |
| MM: MILITARY MEDAL | | | | | |
| SERGEANT | | | | | |
| 39273 | 29.06.51 | 2720659 | ROWAN | Frederick | |
| CORPORAL | | | | | |
| 39528 | 29.04.52 | 22288370 | JENNINGS | Douglas Albert | |
| TROOPER | | | | | |
| 39282 | 10.07.51 | 82323 | BOMBER | Harry Lionel | |

| London Gazette issue No. | London Gazette date | Personal Service number | Name | Christian names | Remarks |
|---|---|---|---|---|---|
| MENTION IN DESPATCHES (FOR GALLANT AND DISTINGUISHED SERVICES - POSTHUMOUS) | | | | | |
| BRONZE OAK LEAF EMBLEM | | | | | |
| CAPTAIN | | | | | |
| 39242 | 01.06.52 | 128778 | ASTLEY-COOPER | D.L. | |
| MENTION IN DESPATCHES (FOR GALLANT AND DISTINGUISHED SERVICES) | | | | | |
| BRONZE OAK LEAF EMBLEM | | | | | |
| LIEUTENANT-COLONEL | | | | | |
| 39328 | 07.09.51 | 53672 | LOWTHER, Bt. | Sir William Guy | |
| CAPTAINS | | | | | |
| 39528 | 29.04.52 | 369050 | LIDSEY | J.H. | King's Dragoon Guards attached |
| 39528 | 29.04.52 | 190453 | MEADE | C.J.G. | |
| LIEUTENANTS | | | | | |
| 39282 | 10.07.52 | 373729 | DOWLING | B.L. | |
| 39528 | 29.04.52 | 403482 | HART | K.G.I. | 7th Queen's Own Hussars attached |
| 39273 | 29.06.51 | 323532 | RADFORD | M.A. | |
| SERGEANTS | | | | | |
| 39528 | 29.04.52 | 7888569 | ADAMS | R.W. | |
| 39528 | 29.04.52 | 4691702 | MAPALS | L. | |
| 39528 | 29.04.52 | 14447848 | MAW | W.J. | |
| LANCE-CORPORAL | | | | | |
| 39528 | 29.04.52 | 22275723 | CHILDS | D.T. | |

| London Gazette issue No. | London Gazette date | Personal Service number | Name | Christian names | Remarks |
|---|---|---|---|---|---|
| ROYAL TANK REGIMENT | | | | | raised 1917 |
| OBE: OFFICER OF THE ORDER OF THE BRITISH EMPIRE | | | | | |
| LIEUTENANT-COLONEL | | | | | |
| 40036 | 04.12.53 | 47586 | HOPKINSON, DSO, MC | Gerald Charles | |
| MBE: MEMBER OF THE ORDER OF THE BRITISH EMPIRE | | | | | |
| MAJOR | | | | | |
| 40036 | 04.12.53 | 338800 | KENNEDY | John Alastair | |
| CAPTAIN | | | | | |
| 40036 | 04.12.53 | 304676 | AMBIDGE, MC | Douglas William Arthur | |
| MM: MILITARY MEDAL | | | | | |
| SERGEANTS | | | | | |
| 40036 | 04.12.53 | 1779557 | BRUNDISH | Frederick James | |
| 40036 | 04.12.53 | 14181341 | MacFARLANE | John | |
| 40036 | 04.12.53 | 14051827 | WALLACE | Allan James George | |
| MENTION IN DESPATCHES (FOR GALLANT AND DISTINGUISHED SERVICES) | | | | | |
| BRONZE OAK LEAF EMBLEM | | | | | |
| MAJORS | | | | | |
| 39906 | 07.07.53 | 162125 | SULLIVAN | W.D.P. | |
| 40036 | 04.12.53 | 73199 | WARD, DSO, MC | R.E. | |
| CAPTAIN | | | | | |
| 39906 | 07.07.53 | 369522 | WELCH | T.S.M. | |
| LIEUTENANTS | | | | | |
| 40036 | 04.12.53 | 407819 | FARMER | M.G. | |
| 39906 | 07.07.53 | 409066 | ULOTH | A.C. | |
| CORPORALS | | | | | |
| 39870 | 01.06.53 | 14194006 | COWLING | J.E. | |
| 39906 | 07.07.53 | 22525828 | GRIFFITHS | G. | 10th Royal Hussars attached |

Those listed in the category Royal Armoured Corps have been so placed because London Gazette entries for them only show parent regiment titles and not those of the regiments they were attached to during the Korean conflict.

| London Gazette issue No. | London Gazette date | Personal Service number | Name | Christian names | Remarks |
|---|---|---|---|---|---|
| ROYAL ARMOURED CORPS | | | | | |
| MC: MILITARY CROSS | | | | | |
| 2ND LIEUTENANT | | | | | |
| 39282 | 10.07.51 | 403139 | VENNER | John Boldrewood | 7th Queen's Own Hussars |
| MENTION IN DESPATCHES (FOR GALLANT AND DISTINGUISHED SERVICES) | | | | | |
| BRONZE OAK LEAF EMBLEM | | | | | |
| MAJOR | | | | | |
| 40036 | 04.12.53 | 71872 | FLETCHER, MBE | J. | 16th/5th The Queen's Royal Lancers |
| CAPTAIN | | | | | |
| 40036 | 04.12.53 | 255837 | WASHINGTON | T.J.C. | 12th Royal Lancers |
| LIEUTENANT | | | | | |
| 39273 | 29.06.51 | 364199 | WOLFE | G.D.H. | 9th Queen's Royal Lancers |
| CORPORAL | | | | | |
| 39906 | 07.07.53 | 22525828 | GRIFFITHS | G. | 10th Royal Lancers |

# ROYAL REGIMENT OF ARTILLERY

| London Gazette issue No. | London Gazette date | Personal Service number | Name | Christian names | Remarks |
|---|---|---|---|---|---|
| ROYAL REGIMENT OF ARTILLERY | | | | | The Gunners; The Right of the Line |
| DSO: DISTINGUISHED SERVICE ORDER | | | | | Raised 1716 |
| LIEUTENANT-COLONEL | | | | | |
| 40036 | 04.12.53 | 50228 | BRENNAN, CBE | Thomas Geoffrey | Commander 20th Field Regiment |
| MAJOR | | | | | |
| 39831 | 21.04.53 | 165454 | PONT | Reginald Albert | Unit not listed |
| OBE: OFFICERS OF THE ORDER OF THE BRITISH EMPIRE | | | | | |
| LIEUTENANT-COLONELS | | | | | |
| 40206 | 15.06.54 | 174809 | De CENT | Douglas Cecil | Unit not listed |
| 39666 | 10.10.52 | 124314 | CALVERT, MC | Henry Stanton | Commander 61st Light Regiment |
| 39666 | 10.10.52 | 41198 | SLADE-POWELL, DSO | Jack Harold | Commander 14th Field Regiment |
| 39528 | 29.04.52 | 41188 | YOUNG, DSO | Maris Theo | Commander 45th Field Regiment |
| MAJORS | | | | | |
| 40036 | 04.12.53 | 284001 | BATTEN, DCM | Arthur John | Unit not listed |
| 39870 | 26.05.53 | 380032 | CABLE, MC | Desmond James | Unit not listed |
| 39528 | 29.04.52 | 158257 | VICTORY, MC | Patrick Michael | Unit not listed |
| MBE: MEMBER OF THE ORDER OF THE BRITISH EMPIRE | | | | | |
| MAJORS | | | | | |
| 40036 | 04.12.53 | 134778 | BULL | Walter | Unit not listed |
| 40036 | 04.12.53 | 74532 | FLETCHER | John Antony | Unit not listed |
| 39666 | 10.10.52 | 95066 | GREGG | John Anthony | Unit not listed |
| 40036 | 04.12.53 | 85547 | JONES | John Henry Peyton | Unit not listed |
| CAPTAINS | | | | | |
| 40206 | 18.06.54 | 189466 | BUTLER | Ronald Austen Dunstan | Unit not listed |
| 39831 | 21.04.53 | 269664 | HEATON | Basil Hugh Philipse | Unit not listed |
| 39831 | 21.04.53 | 148315 | STOTT | Kenneth Frederick | Unit not listed |
| 39831 | 21.04.53 | 358638 | TURNER | Michael Henry | Unit not listed |
| 39870 | 01.06.53 | 245083 | WOODLEY | Hector | Unit not listed |
| CAPTAIN QUARTERMASTER | | | | | |
| 39870 | 01.06.53 | 405399 | HARTLAND | Joseph Phillip | Unit not listed |
| WARRANT OFFICER CLASS I | | | | | |
| 39831 | 21.04.53 | 828231 | HARTNEY | James Francis | Unit not listed |
| 39528 | 29.04.52 | 843737 | SMITH | Raymond George | Unit not listed |
| 39666 | 10.10.52 | 762193 | WALKINGSHAW | Gideon | Unit not listed |
| WARRANT OFFICER CLASS II | | | | | |
| 39666 | 10.10.52 | 784230 | COLE | Cecil John | Unit not listed |
| 40036 | 04.12.53 | 797469 | HUNT | Alfred Leonard | Unit not listed |

| London Gazette issue No. | London Gazette date | Personal Service number | Name | Christian names | Remarks |
|---|---|---|---|---|---|
| BAR TO MC: MILITARY CROSS | | | | | |
| MAJOR | | | | | |
| 40036 | 04.12.53 | 380040 | MACKAY, DSO, MC | William Miller | ← Unit not listed |
| CAPTAIN | | | | | |
| 39641 | 09.09.52 | 158260 | BAXTER, MC | James Arthur Cruice | ← |
| MC: MILITARY CROSS | | | | | |
| MAJORS | | | | | |
| 40036 | 04.12.53 | 278873 | KING-MARTIN | John Douglas | ← |
| 40036 | 04.12.53 | 71008 | SHORE | Francis Scott Gordon | OC 107 Field Battery |
| CAPTAINS | | | | | |
| 40036 | 04.12.53 | 376220 | CHILDS | Peter Desmond Robin | ← |
| 39813 | 31.03.53 | 369870 | DANSKIN | John Chadwick | ← |
| 39666 | 10.10.52 | 335389 | de GREY | Hon. John | ← |
| 39328 | 07.09.51 | 200496 | FOWLE | Anthony Peter Hedley Bruce | ← |
| 39906 | 07.07.53 | 210509 | GORDON | John Lionel Hugh | ← |
| 40036 | 04.12.53 | 296006 | GRANT | Noel Christopher Ogilvie | ← |
| 40036 | 04.12.53 | 289691 | HARRISON | Jerrold Alexander | ← |
| | | 182304 | MORGAN | J.G. | ← |
| 39406 | 11.12.51 | 349773 | STRICKLAND | Gerald John | ← |
| 39528 | 29.04.52 | 149047 | WELDON | Patrick Langford Daryl | 45 Field Regiment, 116 Fd Battery |
| LIEUTENANTS | | | | | |
| 39831 | 24.04.53 | 397914 | HANNA | Robert Houston | ← |
| 39273 | 29.06.51 | 397300 | LAKES | Gordon Harry | ← |
| 39528 | 29.04.52 | 397993 | MONILAWS | Richard Adrian | ← |
| 39273 | 29.06.51 | 360738 | WALSH | Edward Paul | ← |
| 39831 | 24.04.53 | 385048 | WOOD | Geoffrey Alan | ← |
| GM: GEORGE MEDAL | | | | | |
| (IN RECOGNITION OF GALLANT AND DISTINGUISHED SERVICE WHILST A PRISONER OF WAR IN NORTH KOREA) | | | | | |
| CAPTAIN | | | | | |
| 40146 | 09.04.54 | 304047 | GIBBON | Acton Henry Gordon | ← |
| GM: GEORGE MEDAL | | | | | |
| LIEUTENANT | | | | | |
| 39756 | 20.01.53 | 307721 | MAWSON | Beverley Anthony Foster | G Troop 61st Light Regiment |

← Unit not listed

| London Gazette issue No. | London Gazette date | Personal Service number | Name | Christian names | Remarks |
|---|---|---|---|---|---|
| **MM: MILITARY MEDAL** | | | | | |
| **WARRANT OFFICERS CLASS II** | | | | | |
| 40036 | 04.12.53 | 800936 | ASKEW | George Eric | ← Unit not listed |
| 39273 | 29.06.51 | 847297 | ELLIS | Andrew Murray | ← |
| **SERGEANTS** | | | | | |
| 39822 | 10.04.53 | 14188906 | ONYETT | Croydon | ← |
| 40036 | 04.12.53 | 22208199 | SIMPSON | John Boyle | ← |
| **BOMBARDIERS** | | | | | |
| 39870 | 01.06.53 | 22178570 | LEWIS | Eric | ← |
| 40036 | 04.12.53 | 1151574 | NEWBOLD | Arthur | ← |
| **LANCE-BOMBARDIER** | | | | | |
| 40036 | 04.12.53 | 22410094 | MACHEN | Geoffrey | ← |
| **GUNNERS** | | | | | |
| 40036 | 04.12.53 | 22305965 | BOND | Thomas | ← |
| 40036 | 04.12.53 | 22615533 | ELCOAT | William George | ← |
| 39205 | 17.04.51 | 14457810 | SPRAGGS | Herbert James | ← |
| 39282 | 10.07.51 | 926594 | YOUNG | Samuel George | ← |
| **BEM: BRITISH EMPIRE MEDAL** | | | | | |
| **WARRANT OFFICERS CLASS II** | | | | | |
| 40036 | 04.12.53 | 14821977 | ANDERSON | Charles Hutchison | ← |
| 39528 | 29.04.52 | 22154142 | WHITEHEAD | Leslie Addinall | ← |
| **MENTION IN DESPATCHES (FOR GALLANT AND DISTINGUISHED SERVICES - POSTHUMOUS)** | | | | | |
| **BRONZE OAK LEAF EMBLEM** | | | | | |
| **CAPTAINS** | | | | | |
| 40206 | 18.06.54 | 373004 | HOLMAN, MBE | W.M. | ← |
| 40036 | 08.12.53 | 261440 | WASHBROOK | R.F. | ←listed: later died as a POW |
| **MENTION IN DESPATCHES (FOR GALLANT AND DISTINGUISHED SERVICES WHILST A PRISONER OF WAR IN NORTH KOREA - POSTHUMOUS)** | | | | | |
| **BRONZE OAK LEAF EMBLEM** | | | | | |
| **CAPTAIN** | | | | | |
| 40146 | 13.04.54 | 261440 | WASHBROOK | R.F. | ← listed: died as a prisoner of war 21.11.51. |
| **MENTION IN DESPATCHES (FOR GALLANT AND DISTINGUISHED SERVICES)** | | | | | |
| **BRONZE OAK LEAF EMBLEM** | | | | | |
| **LIEUTENANT-COLONEL** | | | | | |
| 39666 | 10.10.52 | 380613 | BICKLEY | G.R.G. | ← |
| | | | | | ← Unit not listed |

| London Gazette issue No. | London Gazette date | Personal Service number | Name | Christian names | Remarks |
|---|---|---|---|---|---|
| **MENTION IN DESPATCHES (FOR GALLANT AND DISTINGUISHED SERVICES) (continued)** | | | | | |
| **BRONZE OAK LEAF EMBLEM** | | | | | |
| **MAJORS** | | | | | |
| 39831 | 24.04.53 | 293280 | CRONIN | K.R. | ↚ Unit not listed |
| 40036 | 04.12.53 | 118412 | DAIN | C.S.R. | ↚ |
| 39528 | 29.04.52 | 58033 | FAWKES, DSO, MC | L.V.F. | Commander 11th (Sphinx) Indep Light AA |
| 39528 | 29.04.52 | 64008 | FISHER-HOCH | T.V. | Commander 170 Indep Mortar Battery |
| 39273 | 29.06.51 | 271491 | GOWER | R.N.L. | ↚ |
| 40036 | 04.12.53 | 67028 | HULEATT | H.G. | ↚ |
| 40036 | 04.12.53 | 103352 | MACREA-BROWN | D.J. | ↚ |
| 39282 | 10.07.51 | 232680 | NICHOLLS | J.E. | 45 Field Regiment, 70 Fd Battery |
| 39484 | 04.03.52 | 165454 | PONT | R.A. | ↚ |
| 39328 | 07.0 .51 | 72569 | SKUSE | L.E.E. | ↚ |
| 39666 | 10.10.52 | 62503 | THOMAS | E.V. | ↚ |
| 40036 | 04.12.53 | 69055 | THOMPSON | C.P. | ↚ |
| 39666 | 10.10.52 | 117053 | WILKES, MC | L.G. | Commander 120th Light AA Battery |
| 39831 | 24.04.53 | 97240 | WILSON, DSO, TD | R. | ↚ |
| **CAPTAINS** | | | | | |
| 40036 | 04.12.53 | 379624 | BIRILES | G.S. | ↚ |
| 40036 | 04.12.53 | 302802 | McDONALD | D. | ↚ |
| 40036 | 04.12.53 | 109172 | McGHEE | D.W. | ↚ |
| 39328 | 07.09.51 | 288992 | MATTHEWS | F.C. | 45 Field Regiment, 176 Fd Battery |
| 40036 | 04.12.53 | 182304 | MORGAN, MC | J.G. | ↚ |
| 40036 | 04.12.53 | 165877 | NEWCOMBE, MC | A.M.L. | ↚ |
| 40036 | 04.12.53 | 354222 | RYAN | D.F. | ↚ |
| 39831 | 24.04.53 | | TEMPLEMAN | B.J. | ↚ |
| 39528 | 29.04.52 | 249049 | THORNELOE | J.G. | ↚ |
| 39205 | 17.04.51 | 96726 | WATERS, TD | A.G.C.M. | ↚ |
| 39328 | 07.09.51 | 149047 | WELDON | P.L.D. | 45 Field Regiment, 116 Fd Battery |
| 40036 | 04.12.53 | 380136 | WISBEY, MC | F.R. | ↚ |
| **LIEUTENANT-QUARTERMASTER** | | | | | |
| 39831 | 24.04.53 | 420452 | JONES | J.F.D. | ↚ |
| **LIEUTENANTS** | | | | | |
| 39273 | 29.06.51 | 380433 | POWELL | P.A. | ↚ |
| 39831 | 24.04.53 | 393276 | RUSSELL | E.J.U. | ↚ |
| **2ND LIEUTENANT** | | | | | |
| 40036 | 04.12.53 | 423141 | GODFREY | H.W. | ↚ |
| **WARRANT OFFICER CLASS I** | | | | | |
| 39831 | 24.04.53 | 798068 | STEWART | G.K. | ↚ |

↚ Unit not listed

| London Gazette issue No. | London Gazette date | Personal Service number | Name | Christian names | Remarks |
|---|---|---|---|---|---|
| MENTION IN DESPATCHES (FOR GALLANT AND DISTINGUISHED SERVICES) (continued) | | | | | |
| BRONZE OAK LEAF EMBLEM | | | | | |
| BATTERY QUARTERMASTER SERGEANT | | | | | |
| 39531 | 02.05.52 | 22535175 | SIDDLE | J.W. | ∢ Unit not listed |
| WARRANT OFFICERS CLASS II | | | | | |
| 39328 | 07.09.51 | 548910 | DAVENPORT | F.J. | ∢ |
| 39831 | 24.04.53 | 872775 | DAVIES | S.T. | ∢ |
| 39666 | 10.10.52 | 821809 | FENNER | J.G. | ∢ |
| 39282 | 10.07.52 | 847062 | HORSELL | A.A. | ∢ |
| 39528 | 29.04.52 | 861886 | McDONALD | J.P. | ∢ |
| 39870 | 01.06.53 | 835746 | RIDDLE | W.F.J. | ∢ |
| 40036 | 04.12.53 | 14476034 | TALBOT | R.C. | ∢ |
| 40036 | 04.12.53 | 808401 | TIGHE | W. | ∢ |
| 40036 | 04.12.53 | 856391 | WELLS | A.A. | ∢ |
| 39831 | 24.04.53 | 826081 | WICK | E.F. | ∢ |
| 39666 | 10.10.52 | 1526220 | WOOD | G.T.S. | ∢ |
| STAFF-SERGEANT | | | | | |
| 39666 | 10.10.52 | 890262 | CANNON | V.C. | ∢ |
| SERGEANTS | | | | | |
| 39861 | 24.04.53 | 880226 | CHAPPLE | P.H. | ∢ |
| 39328 | 07.09.51 | 845606 | HALES | J.D. | ∢ |
| 39574 | 17.06.52 | 22518404 | JOYCE | M. | ∢ |
| 40036 | 04.12.53 | 1151760 | OLDHAM | D.J. | ∢ |
| 39666 | 10.10.52 | 22275741 | SAINSBURY | W.J. | ∢ |
| 40036 | 04.12.53 | 942043 | SMITH | R.A.J. | ∢ |
| 39666 | 10.10.52 | 19200056 | THOMAS | I.H.T. | ∢ |
| 39831 | 24.04.53 | 1157354 | TITHERIDGE | S.R. | ∢ |
| 39666 | 10.10.52 | 889402 | WILLEY | D.T. | ∢ |
| LANCE-BOMBARDIERS | | | | | |
| 40036 | 04.12.53 | 22666529 | DIXON | P. | ∢ |
| 40036 | 04.12.53 | 22618457 | McCORMACK | A.N. | ∢ |
| 39870 | 01.06.53 | 22233184 | SEARLE | L.J.T. | ∢ |
| 39870 | 01.06.53 | 22522179 | WALSH | E. | ∢ |
| 40036 | 04.12.53 | 22462345 | WOOD | J.H.K. | ∢ |

∢ Unit not listed

| London Gazette issue No. | London Gazette date | Personal Service number | Name | Christian names | Remarks |
|---|---|---|---|---|---|
| MENTION IN DESPATCHES (FOR GALLANT AND DISTINGUISHED SERVICES) (continued) | | | | | |
| BRONZE OAK LEAF EMBLEM | | | | | |
| GUNNERS | | | | | |
| 39666 | 10.10.52 | 22417937 | BERRY | G. | ← Unit not listed |
| 39273 | 29.06.51 | 1469809 | CRISP | A.E. | ← |
| 29682 | 28.10.52 | 22285114 | CROFT | I. | ← |
| 39205 | 17.04.51 | 1103192 | LAW | P.Y. | ← |
| 39288 | 17.07.51 | 14465859 | LOWRY | G. | 170 Independent Mortar Battery |
| 39641 | 09.09.52 | 22476955 | PARTRIDGE | M. | ← |
| 39205 | 17.04.51 | 886950 | STOWE | E.S. | ← |
| 39528 | 29.04.52 | 22400583 | STRONG | G.W. | ← |

## AWARDS CONFERRED BY THE PRESIDENT OF THE UNITED STATES OF AMERICA

| London Gazette issue No. | London Gazette date | Personal Service number | Name | Christian names | Remarks |
|---|---|---|---|---|---|
| SILVER STAR | | | | | |
| LIEUTENANT-COLONEL | | | | | |
| 39462 | 08.02.52 | 41188 | YOUNG, DSO | Maris Theo | Commander 45 Field Regiment |
| MAJORS | | | | | |
| 39462 | 08.02.52 | 64008 | FISHER-HOCH | Terance Vincent | Commander 170 Indep Mortar Battery |
| 29462 | 08.02.52 | 47651 | WITHERS, MC | Henry Clements | 2 I/C 45 Field Regiment |
| BRONZE STAR WITH 'V' DEVICE | | | | | |
| GUNNER | | | | | |
| 39999 | 27.10.53 | 22464504 | SOANES | Peter | ← |
| BRONZE STAR | | | | | |
| MAJOR | | | | | |
| 39999 | 27.10.53 | 62503 | THOMAS | Evan Vaughan | ← |
| CAPTAINS | | | | | |
| 39254 | 08.06.51 | 106695 | O'FLAHERTY, DSO | Denis William Venables Patrick | 45 Field Regiment, 116 Fd Battery |
| 39999 | 27.10.53 | 158257 | VICTORY, OBE, MC | Patrick Michael | ← |

← Unit not listed

| Personal Service number | Name | Christian names | Remarks |
|---|---|---|---|

FRENCH AWARDS

ROYAL REGIMENT OF ARTILLERY

CROIX DE GUERRE DES THÉÂTRES D'OPERATIONS

EXTÉRIEURS AVEC ÉTOILE DE BRONZE

| Personal Service number | Name | Christian names | Remarks |
|---|---|---|---|
| **BRIGADIER** | | | |
| 34436 | GREGSON, DSO, MC | Guy Patrick | C.R.A. 1st Comwel Division |
| **LIEUTENANT-COLONEL** | | | |
| 50228 | BRENNAN, CBE, DSO | Thomas Geoffrey | Officer Commanding 20th Field Regiment, RA |
| **MAJOR** | | | |
| 71008 | SHORE, MC | Francis Scott Gordon | 107 Field Battery, 20th Field Regiment, RA |
| **CAPTAINS** 182304 | | | |
| 182304 | MORGAN, MC | John Geoffrey | 107 Field Battery, 20th Field Regiment, RA |
| 380590 | FAIRGRIEVE | Hugh | 107 Field Battery, 20th Field Regiment, RA |

Order No.108
Lieutenant-Colonel De Germiny, Commanding Officer, Bataillon Français de l'ONU en Corée, 19th October, 1953.

Editor's note: for reasons which are not known, these awards were not acknowledged by the British Authorities in 1953 but, through the efforts of The British Korean Veterans Association and The Association Nationale des Anciens des Forces Françaises de l'O.N.U. et du Regiment de Corée, they were presented to the officers in 1991, with the exception of Major-General Gregson. His widow received the medal.

CORPS OF ROYAL ENGINEERS

ROYAL CORPS OF SIGNALS

| London Gazette issue No. | London Gazette date | Personal Service number | Name | Christian names | Remarks |
|---|---|---|---|---|---|
| CORPS OF ROYAL ENGINEERS | | | | | The Sappers: raised 1717 |
| 2ND BAR TO DSO: DISTINGUISHED SERVICE ORDER | | | | | |
| LIEUTENANT-COLONEL | | | | | |
| 39666 | 10.10.52 | 52674 | MOORE, DSO, MC | Peter Neil Martin | Commander 28th Field Engineer Regiment |
| DSO: DISTINGUISHED SERVICE ORDER | | | | | |
| MAJOR | | | | | |
| 39666 | 10.10.52 | 72490 | FLETCHER | Derek George Murchison | ← |
| OBE: OFFICER OF THE ORDER OF THE BRITISH EMPIRE | | | | | |
| LIEUTENANT-COLONEL | | | | | |
| 40036 | 04.12.53 | 47560 | FIELD, MC | Arthur Maurice | Commander 28th Field Engineer Regiment |
| MBE: MEMBER OF THE ORDER OF THE BRITISH EMPIRE | | | | | |
| MAJORS | | | | | |
| 39870 | 01.06.53 | 66629 | GRICE | James Dickenson | ← |
| 39831 | 21.04.53 | 131732 | SHILLAND | Alexander Turner | ← |
| 39666 | 10.10.52 | 346784 | STEADMAN | Terence William Aubrey | ← |
| 39666 | 10.10.52 | 71873 | WOODS, MC | Charles William | ← |
| 40036 | 04.12.53 | 300503 | LEIGH | Ian Aldan Poyntz-Gaynor | Commander 64th Field Park Squadron |
| CAPTAIN | | | | | |
| 40036 | 04.12.53 | 342238 | TURNER | Edward William | ← |
| WARRANT OFFICER CLASS I | | | | | |
| 39666 | 10.10.52 | 1869412 | HACKER | Sydney | ← |
| WARRANT OFFICER CLASS II | | | | | |
| 39831 | 21.04.53 | 1882851 | HIRST | Kenneth | ← |
| BAR TO MC: MILITARY CROSS | | | | | |
| CAPTAIN | | | | | |
| 39282 | 10.07.51 | 288423 | BAYTON-EVANS, MC | Herbert | ← |
| MC: MILITARY CROSS | | | | | |
| MAJORS | | | | | |
| 39831 | 24.04.53 | 273539 | FROSNELL | Serge Albert | ← |
| 40036 | 04.12.53 | 125185 | HANNAY | Vivian Henry Spencer | ← |
| CAPTAINS | | | | | |
| 39870 | 01.06.53 | 364703 | CAMERON | David Ewan Reeves | ← |

← Unit not listed

| London Gazette issue No. | London Gazette date | Personal Service number | Name | Christian names | Remarks |
|---|---|---|---|---|---|
| MC: MILITARY CROSS (continued) | | | | | |
| CAPTAINS (continued) | | | | | |
| 39574 | 17.06.52 | 362289 | CARR | Colin Dening | ∢ Unit not listed |
| 40036 | 04.12.53 | 357063 | COOPER | George Leslie Conroy | ∢ |
| 39666 | 10.10.52 | 393107 | CORMACK | John Napier | ∢ |
| 39273 | 29.06.51 | 195038 | HOLMES | William Desmond Cuthbert | ∢ |
| 39666 | 10.10.52 | 352532 | JAMES | Alan Christopher | ∢ |
| 39528 | 29.04.52 | 152684 | PAGE | John Humphrey | ∢ |
| 39765 | 30.01.53 | 382620 | SULLIVAN | Richard Terence Derek | ∢ |
| 2ND LIEUTENANTS | | | | | |
| 39528 | 29.04.52 | 413098 | PALMER | Trevor | ∢ |
| 39831 | 24.04.53 | 418413 | THOMPSON | Ian Alexander Dorling | ∢ |
| DCM: DISTINGUISHED CONDUCT MEDAL | | | | | |
| SERGEANT | | | | | |
| 39282 | 10.07.51 | 1880080 | ORTON | Reynold Arthur | ∢ |
| MM: MILITARY MEDAL | | | | | |
| SERGEANTS | | | | | |
| 39666 | 10.10.52 | 14463448 | NEAVE | Leslie Arthur | ∢ |
| 39528 | 29.04.52 | 19037937 | RANKIN | Hugh Stewart | ∢ |
| CORPORALS | | | | | |
| 39682 | 28.10.52 | 22523482 | FORD | Leonard Ernest | ∢ |
| 39831 | 24.04.53 | 22541338 | FOX | Willis | ∢ |
| 39788 | 27.02.53 | 22550560 | JENKINS | Ievan Cynwyd | ∢ |
| 39528 | 27.02.53 | 22236996 | WEAVER | Albert | ∢ |
| LANCE-CORPORAL | | | | | |
| 40036 | 04.12.53 | 22637355 | GRIFFITHS | Ronald Henry | ∢ |
| SAPPER | | | | | |
| 39906 | 07.07.53 | 22308151 | SMYTH | John | ∢ |
| BEM: BRITISH EMPIRE MEDAL | | | | | |
| WARRANT OFFICER CLASS II | | | | | |
| 39831 | 21.04.53 | 1874697 | JONES | Llewelyn George | ∢ |
| STAFF-SERGEANT | | | | | |
| 39666 | 10.10.52 | 22211604 | GOLDSMITH | David Edward | ∢ |

∢ Unit not listed

| London Gazette issue No. | London Gazette date | Personal Service number | Name | Christian names | Remarks |
|---|---|---|---|---|---|
| MENTION IN DESPATCHES (FOR GALLANT AND DISTINGUISHED SERVICES) | | | | | |
| BRONZE OAK LEAF EMBLEM | | | | | |
| MAJORS | | | | | |
| 39328 | 07.09.51 | 322888 | BARBER | W.B. | € Unit not listed |
| 39328 | 07.09.51 | 121153 | DURHAM, MBE | R.H.P.L. | € |
| 39666 | 10.10.52 | 95181 | KEER | J.A. | Commander 64th Field Park Squadron |
| 39831 | 24.04.53 | 201709 | LESLIE, MC | P. | € |
| 39528 | 29.04.52 | 95184 | STEPHENS, MBE | H.W.B. | € |
| CAPTAINS | | | | | |
| 40036 | 04.12.53 | 382622 | BALL | P.W. | € |
| 39528 | 29.04.52 | 229336 | BEAN | K.M. | € |
| 40036 | 04.12.53 | 281279 | GLADWYN, MBE | R.F.C. | Royal Corps of Signals attached RE |
| 40036 | 04.12.53 | 339962 | GRAHAM | R.E. | € |
| 39666 | 10.10.52 | 368293 | McKAY-FORBES | J.A. | € |
| 39831 | 24.04.53 | 242484 | POULTON | J.W.L. | € |
| 40036 | 04.12.53 | 253640 | RICHARDS | T.L. | € |
| 39870 | 01.06.53 | 368784 | WATLING | T.F. | € |
| LIEUTENANTS | | | | | |
| 39273 | 29.06.51 | 397200 | CADOUX-HUDSON | D. | € |
| 40036 | 04.12.53 | 397870 | DUDDY | A. | € |
| 40036 | 04.12.53 | 383927 | HARDHAM | J.A. | € |
| 40036 | 04.12.53 | 415980 | MOORES | P.R. | € |
| 2ND LIEUTENANT | | | | | |
| 40036 | 04.12.53 | 423888 | HEATON | P. | € |
| WARRANT OFFICERS CLASS II | | | | | |
| 39831 | 24.04.53 | 1878132 | STRATFORD | C.C. | € |
| 39666 | 10.10.52 | 22282092 | TARRANT | C. | € |
| STAFF-SERGEANT | | | | | |
| 40036 | 04.12.53 | 14623262 | HOLDING | J.H. | € |
| SERGEANTS | | | | | |
| 39666 | 10.10.52 | 22521579 | ATTWOOD | R.H.B. | € |
| 39528 | 29.04.52 | 14463488 | NEAVE | L.A. | € |
| 39666 | 10.10.52 | 14440016 | OWEN | N.D. | € |
| 39328 | 07.09.51 | 876651 | PHELPS | N.D. | € |
| 39870 | 01.06.53 | 1878337 | RICHARDSON | M.G. | € |
| 39870 | 01.06.53 | 22541733 | THOMPSON | R.I. | € |
| 40036 | 04.12.53 | 14033425 | WILLIAMS | T.J. | € |
| 39831 | 24.04.53 | 14185164 | YATES | L. | € |

€ Unit not listed

| London Gazette issue No. | London Gazette Date | Personal Service number | Name | Christian names | Remarks |
|---|---|---|---|---|---|
| MENTION IN DESPATCHES (FOR GALLANT AND DISTINGUISHED SERVICES) (continued) | | | | | |
| CORPORALS | | | | | |
| 39870 | 01.06.53 | 22494356 | BEELS | T. | ← Unit not listed |
| 40036 | 04.12.53 | 21126308 | HOWELL | J.G. | ← |
| 39666 | 10.10.52 | 14469098 | MALLOY | E.F. | ← |
| 40036 | 04.12.53 | 22307230 | NORTON | L. | ← |
| 40036 | 04.12.53 | 22540067 | SIMMONDS | H.G. | ← |
| 39661 | 03.10.52 | 22273531 | WELLS | B. | ← |
| LANCE-CORPORALS | | | | | |
| 39528 | 29.04.52 | 22521585 | EVANS | P.A. | ← |
| 39831 | 24.04.53 | 22507973 | MURRAY | J.W.F. | ← |
| 39661 | 03.10.52 | 22274801 | WALL | R.F.G. | ← |
| 39528 | 29.04.52 | 2239590 | WOODFORD | I.A. | ← |
| SAPPERS | | | | | |
| 40036 | 04.12.53 | 22587684 | CROOK | D. | ← |
| 39528 | 29.04.52 | 21181275 | JOLLY | G.R. | ← |
| 39870 | 01.06.53 | 22309688 | MILLER | A. | ← |
| 39528 | 29.04.52 | 1878388 | RUSHBROOK | A.E.M. | ← |
| 39666 | 10.10.52 | 22249051 | SINCLAIR | J. | ← |
| 39273 | 29.06.51 | 22286057 | WATSON | J. | ← |

## AWARDS CONFERRED BY THE PRESIDENT OF THE UNITED STATES OF AMERICA

| London Gazette issue No. | London Gazette Date | Personal Service number | Name | Christian names | Remarks |
|---|---|---|---|---|---|
| SILVER STAR | | | | | |
| MAJOR | | | | | |
| 39462 | 08.02.52 | 85568 | YOUNGER, DSO, BA | Allan Elton | Commander 55th Field Squadron |
| BRONZE STAR | | | | | |
| CAPTAIN | | | | | |
| 39999 | 27.10.53 | 228852 | KEER, MC | Martin Cordy | Commander 64th Field Park Squadron |
| CORPORAL | | | | | |
| 39254 | 08.06.51 | 1976543 | CROSS | Gerald Norman | ← |
| SAPPERS | | | | | |
| 39254 | 08.06.51 | 14417553 | HANNON | James Christopher | ← |
| 39254 | 08.06.51 | 22249973 | PRESTON | Alfred Frank John | ← |

← Unit not listed

| London Gazette issue No. | London Gazette date | Personal Service number | Name | Christian names | Remarks |
|---|---|---|---|---|---|
| ROYAL CORPS OF SIGNALS | | | | | Raised 1920 |
| OBE: OFFICER OF THE ORDER OF THE BRITISH EMPIRE | | | | | |
| LIEUTENANT-COLONELS | | | | | |
| 39528 | 29.04.52 | 62965 | ATKINSON, B.Sc., AMIEE | Arthur Leslie | Commander 1st Comwel Div Signals Regiment |
| 40036 | 04.12.53 | 53657 | BLAKER | Edward Cecil Richard | Commander 1st Comwel Div Signals Regiment |
| 39831 | 21.04.53 | 56630 | JOHNSON, MBE | Frederick Peter | Commander 1st Comwel Div Signals Regiment |
| MBE: MEMBER OF THE ORDER OF THE BRITISH EMPIRE | | | | | |
| MAJORS | | | | | |
| 39870 | 01.06.53 | 237810 | BRACKENBURY | Stanley William | € |
| 39528 | 29.04.52 | 95088 | FENTON | George Symmes | € |
| 39528 | 29.04.52 | 56630 | JOHNSON | Frederick Peter | € |
| 40036 | 04.12.53 | 95201 | SAWERS | James Maxwell | € |
| MAJOR QUARTERMASTER | | | | | |
| 39666 | 10.10.52 | 105596 | NAPPER | Harold | € |
| CAPTAINS | | | | | |
| 40036 | 04.12.53 | 286773 | MASSEY | James Anthony | € |
| 39831 | 21.04.53 | 373207 | WEBB | Patrick Henry Fane | € |
| MM: MILITARY MEDAL | | | | | |
| CORPORAL | | | | | |
| 39282 | 10.07.51 | 1103341 | BESTWICK | Neville | € |
| BEM: BRITISH EMPIRE MEDAL | | | | | |
| WARRANT OFFICERS CLASS II | | | | | |
| 39666 | 10.10.52 | 2320456 | O'FLAHERTY | Michael Joseph | € |
| 40036 | 04.12.53 | 6845001 | PARISH | Alfred | € |
| SERGEANTS | | | | | |
| 39666 | 10.10.52 | 2327553 | CORFIELD | William John | € |
| 39528 | 29.04.52 | 14326820 | SIMPSON | Robert | € |
| MENTION IN DESPATCHES (FOR GALLANT AND DISTINGUISHED SERVICES) | | | | | |
| BRONZE OAK LEAF EMBLEM | | | | | |
| MAJORS | | | | | |
| 39666 | 10.10.52 | 73408 | DENTON | J.E. | € |
| 40036 | 04.12.53 | 391336 | HUMPHRIES | D.M. | € |
| 39328 | 07.09.51 | 57440 | MARSHALL | A.R. | € |

€ Unit not listed

| London Gazette issue No. | London Gazette date | Personal Service number | Name | Christian names | Remarks |
|---|---|---|---|---|---|
| MENTION IN DESPATCHES (FOR GALLANT AND DISTINGUISHED SERVICES) (continued) | | | | | |
| BRONZE OAK LEAF EMBLEM | | | | | |
| CAPTAINS | | | | | |
| 39831 | 24.04.53 | 406347 | BIRTWISTLE | A.C. | € Unit not listed |
| 39831 | 24.04.53 | 382536 | EVANS | P.J. | attached Royal Artillery; Unit not listed |
| 40036 | 04.12.53 | 281279 | GLADWYN, MBE | R.F.C. | attached Corps Royal Engineers; Unit not listed |
| 40036 | 04.12.53 | 210376 | JACKSON | T.G.H | € |
| 39666 | 10.10.52 | 393193 | LANGWORTHY | G.L.N. | € |
| 39666 | 10.10.52 | 227521 | RANSOME | H.N. | € |
| 40036 | 04.12.53 | 253640 | RICHARDS | T.L. | € |
| LIEUTENANTS | | | | | |
| 39328 | 07.09.51 | 403513 | KIRKBY | K. | € |
| 40036 | 04.12.53 | 415980 | MOORES | P.R. | € |
| WARRANT OFFICERS CLASS II | | | | | |
| 39831 | 24.04.53 | 2327711 | LANE | N. | € |
| 39831 | 24.04.53 | 3599642 | SWENSON | D.W. | € |
| SERGEANTS | | | | | |
| 39831 | 24.04.53 | 22096929 | AVEN | G.A. | € |
| 39666 | 10.10.52 | 22265493 | HARDY | F. | € |
| 39666 | 10.10.52 | 14470816 | HUTTON | J.R. | € |
| CORPORALS | | | | | |
| 39870 | 01.06.53 | 22496926 | HARVEY | K.R.J. | € |
| 39666 | 10.10.52 | 22522716 | MERRYWEATHER | G.E. | € |
| 39328 | 07.09.51 | 21005369 | PRINCE | H.E. | € |
| LANCE-CORPORALS | | | | | |
| 39328 | 07.09.51 | 22274460 | HODGETTS | E.J.A. | € |
| 39870 | 01.06.53 | 22272636 | HOLT | S. | € |
| 40036 | 04.12.53 | 22212720 | JENNINGS | H. | € |
| SIGNALMEN | | | | | |
| 39328 | 07.09.51 | 22265515 | BAIN | R.H. | € |
| 39528 | 29.04.52 | 22395724 | CLARK | T.M. | € |
| 39528 | 29.04.52 | 22416911 | KINSELLA | R.J. | € |

€ Unit not listed

| London Gazette issue No. | London Gazette date | Personal Service number | Name | Christian names | Remarks |
|---|---|---|---|---|---|

AWARDS CONFERRED BY THE PRESIDENT OF THE UNITED STATES OF AMERICA

BRONZE STAR

MAJORS

| | | | | | |
|---|---|---|---|---|---|
| 39999 | 27.10.53 | 240335 | DODD | James | € Unit not listed |
| 39999 | 27.10.53 | 278115 | LAMB | John McKay | € |
| 39462 | 08.02.52 | 57440 | MARSHALL, TD | Austin Rhodes | € |

CAPTAIN

| | | | | | |
|---|---|---|---|---|---|
| 39999 | 27.10.53 | 233407 | HAYCOCK | Keith Edward | € |

€ Unit not listed

# INFANTRY REGIMENTS OF FOOT

| London Gazette issue No. | London Gazette date | Personal Service number | Name | Christian names | Remarks |
|---|---|---|---|---|---|
| THE ROYAL NORTHUMBERLAND FUSILIERS | | | | | The Fighting Fifth; Old and Bold raised 1674 |
| GC: GEORGE CROSS | | | | | |
| (FOR GALLANT AND DISTINGUISHED SERVICE WHILST A PRISONER OF WAR IN NORTH KOREA) | | | | | |
| FUSILIER | | | | | |
| 40146 | 13.04.54 | 22105517 | KINNE | Derek Godfrey | |
| DSO: DISTINGUISHED SERVICE ORDER | | | | | |
| LIEUTENANT-COLONEL | | | | | |
| 39285 | 13.07.51 | 04768 | FOSTER, OBE | Kingsley Osborne Nugent | Officer Commanding listed: later killed in action |
| MAJORS | | | | | |
| 39282 | 10.07.51 | 78033 | MITCHELL | Charles Hamilton | |
| 39205 | 17.04.51 | 63571 | PRATT | Reginald Michael | |
| 39273 | 29.06.51 | 58900 | WINN, MC | Henry John | |
| MBE: MEMBER OF THE ORDER OF THE BRITISH EMPIRE | | | | | |
| (FOR GALLANT AND DISTINGUISHED SERVICES WHILST A PRISONER OF WAR IN NORTH KOREA) | | | | | |
| CAPTAIN | | | | | |
| 40206 | 18.06.54 | 189193 | DeQUIDT | John Edward Max | Royal Warwickshire Regiment attached |
| MC: MILITARY CROSS | | | | | |
| MAJORS | | | | | |
| 39398 | 30.11.51 | 95242 | BROOK | Reginald David | |
| 39282 | 10.07.51 | 77604 | LEITH-MACGREGOR, DFC | Robert | |
| 2ND LIEUTENANTS | | | | | |
| 39398 | 30.11.51 | 407730 | ADAMS-ACTON | Leo Samuel | |
| 39160 | 02.03.51 | 400455 | CUBISS | John Malcolm | West Yorkshire Regiment attached |
| 39282 | 10.07.51 | 403682 | SHEPPARD | William Paine | East Yorkshire Regiment attached |
| 39398 | 30.11.51 | 387678 | YEO | James Alfred | East Yorkshire Regiment attached |
| 39282 | 10.07.51 | 403746 | YOUNG | Matthew Dennis | |
| DCM: DISTINGUISHED CONDUCT MEDAL | | | | | |
| WARRANT OFFICER CLASS II (CSM) | | | | | |
| 39282 | 10.07.51 | 4269977 | RADCLIFFE | Edward John | |
| FUSILIER | | | | | |
| 39273 | 29.06.51 | 22525874 | CROOKES, MM | Ronald | |

| London Gazette issue No. | London Gazette date | Personal Service number | Name | Christian names | Remarks |
|---|---|---|---|---|---|
| MM: MILITARY MEDAL | | | | | |
| SERGEANT | | | | | |
| 39205 | 17.04.51 | 4397370 | PILCHER | John | |
| CORPORALS | | | | | |
| 39282 | 10.07.51 | 4390499 | HADFIELD | Lancelot | |
| 39282 | 10.07.51 | 19037566 | THOMPSON | James Dennis | |
| LANCE-CORPORAL | | | | | |
| 39282 | 10.07.51 | 1427036 | OVEN | Joseph | |
| FUSILIERS | | | | | |
| 39205 | 17.04.51 | 4690537 | BARKER | James Baden | |
| 39528 | 29.04.52 | 14467097 | WAPPETT | William | York and Lancaster Regiment attached listed: later died |
| BEM: BRITISH EMPIRE MEDAL | | | | | |
| (FOR GALLANT AND DISTINGUISHED SERVICE WHILST A PRISONER OF WAR IN NORTH KOREA) | | | | | |
| SERGEANT | | | | | |
| 40146 | 13.04.54 | 14472846 | SHARP | David | Last British/Commonwealth POW released |
| MENTION IN DESPATCHES (FOR GALLANT AND DISTINGUISHED SERVICE WHILST A PRISONER OF WAR - POSTHUMOUS) | | | | | |
| BRONZE OAK LEAF EMBLEM | | | | | |
| LIEUTENANT | | | | | |
| 40146 | 09.04.54 | 407730 | ADAMS-ACTON | Leo Samuel | |
| MENTION IN DESPATCHES (FOR GALLANT AND DISTINGUISHED SERVICES) | | | | | |
| BRONZE OAK LEAF EMBLEM | | | | | |
| LIEUTENANT-COLONEL | | | | | |
| 39328 | 07.09.51 | 38606 | SPEER | M.C. | Officer Commanding |
| MAJORS | | | | | |
| 39398 | 30.11.51 | 115013 | KERSHAW | R.M. | |
| 39205 | 17.04.51 | 58900 | WINN, MC | J. | |
| MAJOR QUARTERMASTER | | | | | |
| 39528 | 29.04.52 | 123943 | PURCELL, MBE | J.W.M. | |
| LIEUTENANTS | | | | | |
| 39398 | 30.11.51 | 115013 | PHILLIPS | S.A.S. | |
| 39328 | 07.09.51 | 387678 | YEO | J.A. | East Yorkshire Regiment attached |

| London Gazette issue No. | London Gazette date | Personal Service number | Name | Christian names | Remarks |
|---|---|---|---|---|---|
| MENTION IN DESPATCHES (FOR GALLANT AND DISTINGUISHED SERVICES) (continued) | | | | | |
| BRONZE OAK LEAF EMBLEM | | | | | |
| WARRANT OFFICER CLASS I (RSM) | | | | | |
| 39528 | 29.04.52 | 4269809 | RICHARDSON | G.E. | |
| WARRANT OFFICER CLASS II (CSM) | | | | | |
| 39528 | 29.04.52 | 4269637 | REYNOLDS | P.J. | |
| SERGEANTS | | | | | |
| 39205 | 17.04.51 | 4344818 | BAKER | K.E.M. | |
| 39666 | 10.10.52 | 4343198 | BROWN | G. | |
| 39528 | 29.04.51 | 2266183 | HEPTON | I.A. | |
| CORPORAL | | | | | |
| 39205 | 17.04.51 | 4390494 | CLARKE | W.E. | |
| FUSILIERS | | | | | |
| 39273 | 29.06.51 | 4390084 | BROWN | G.V. | |
| 39398 | 30.11.51 | 22267989 | COCKBURN | G.H.W. | |
| 39273 | 29.06.51 | 4987218 | FEARN | A. | |
| 39528 | 29.04.51 | 51110107 | McCORMACK | O.J. | |
| 39273 | 29.06.51 | 19152416 | STEPHENSON | D. | |
| 39273 | 29.06.51 | 14467097 | WAPPETT | W. | York & Lancaster Regiment attached listed: later died |
| 39398 | 30.11.51 | 14423787 | WHALEY | J. | |
| 39870 | 01.06.53 | 22019052 | WILLIAMS | R. | |

## AWARDS CONFERRED BY THE PRESIDENT OF THE UNITED STATES OF AMERICA

| London Gazette issue No. | London Gazette date | Personal Service number | Name | Christian names | Remarks |
|---|---|---|---|---|---|
| BRONZE STAR | | | | | |
| CAPTAIN | | | | | |
| 39254 | 08.06.51 | 74399 | ELLERY, TD | Hugh William | |
| US ARMY COMMENDATION MEDAL WITH V DEVICE FOR VALOUR | | | | | |
| SERGEANT | | | | | |
| * Order 358 | 17.12.50 | 14472846 | SHARP | David | Field Intelligence attached HQ I Corps US Army, Liaison Section, on strategic operations. HQ I Corps Order No. 358 refers. |

* Last British/Commonwealth POW to be released

| London Gazette issue No. | London Gazette date | Personal Service number | Name | Christian names | Remarks |
|---|---|---|---|---|---|
| THE ROYAL FUSILIERS (CITY OF LONDON REGIMENT) | | | | | Elegant Extracts raised 1685 |
| MBE: MEMBER OF THE ORDER OF THE BRITISH EMPIRE | | | | | |
| MAJOR | | | | | |
| 39870 | 01.06.53 | 64586 | HARDING | Basil Edgerton Mulcaster | |
| MC: MILITARY CROSS | | | | | |
| MAJOR | | | | | |
| 40036 | 04.12.53 | 95246 | HILL | Henry Hubert | |
| LIEUTENANT | | | | | |
| 40166 | 07.05.54 | 400343 | De ROEPER | Peter John | |
| 40036 | 04.12.53 | 408057 | TEAGUE | John Anthony | |
| DCM: DISTINGUISHED CONDUCT MEDAL | | | | | |
| FUSILIER | | | | | |
| 39883 | 09.06.53 | 22508234 | HODKINSON | George Arthur | East Surrey Regiment attached |
| MM: MILITARY MEDAL | | | | | |
| CORPORAL | | | | | |
| 39898 | 26.06.53 | 22597720 | FOX | Major John | |
| LANCE-CORPORAL | | | | | |
| 37959 | 23.01.53 | 19033008 | CLARK | George Arthur | |
| FUSILIER | | | | | |
| 40036 | 04.12.53 | 22612736 | DALTON | John Sidney | Queen's Royal Regiment attached |
| MENTION IN DESPATCHES (FOR GALLANT AND DISTINGUISHED SERVICES - POSTHUMOUS) | | | | | |
| BRONZE OAK LEAF EMBLEM | | | | | |
| 2ND LIEUTENANT | | | | | |
| 39759 | 23.01.53 | 419711 | HOARE | C.R.P.M. | |
| 39898 | 26.06.53 | 423033 | RUHEMANN | R.H. | |
| FUSILIER | | | | | |
| 39898 | 26.06.53 | 22581456 | GWYNNE | S.J. | Queen's Royal Regiment attached |
| MENTION IN DESPATCHES (FOR GALLANT AND DISTINGUISHED SERVICES) | | | | | |
| LIEUTENANT-COLONEL | | | | | |
| 40036 | 04.12.53 | 41437 | STEVENS, OBE | G.R. | Officer Commanding |
| MAJORS | | | | | |
| 40036 | 04.12.53 | 67107 | ARCHER, MBE | G.F.H. | |
| 40036 | 04.12.53 | 134182 | DONNELLY, MBE | T.A. | |
| 40036 | 04.12.53 | 74801 | SCHOFIELD, DSO, TD | D.R. | |

| London Gazette issue No. | London Gazette date | Personal Service number | Name | Christian name | Remarks |
|---|---|---|---|---|---|
| MENTION IN DESPATCHES (FOR GALLANT AND DISTINGUISHED SERVICES) (continued) | | | | | |
| BRONZE OAK LEAF EMBLEM | | | | | |
| CAPTAIN QUARTERMASTER | | | | | |
| 40036 | 04.12.53 | 313019 | BALMER, MBE | H. | |
| LIEUTENANTS | | | | | |
| 39831 | 24.04.53 | 410460 | CONINGTON | J.F. | |
| 40036 | 04.12.53 | 411947 | FRENCH | M.A. | |
| 2ND LIEUTENANTS | | | | | |
| 39759 | 23.01.53 | 420637 | LARRETT | J.H. | |
| 40036 | 04.12.53 | 422255 | PLATT | W.G.C. | |
| WARRANT OFFICER CLASS II | | | | | |
| 40036 | 04.12.53 | 6457389 | HOPWOOD | G.T.D. | |
| SERGEANTS | | | | | |
| 40036 | 04.12.53 | 22548413 | CULLEY | J.C. | |
| 39831 | 24.04.53 | 22517375 | NASH | W.H.M. | |
| FUSILIERS | | | | | |
| 39759 | 23.01.53 | 22508181 | BROMLEY | D.J. | |
| 40036 | 04.12.53 | 22629161 | HOGBIN | J. | The Buffs attached |
| 39666 | 10.10.52 | 22287593 | JONES | W.H. | The Middlesex Regiment attached |
| 39759 | 23.01.53 | 22480799 | MOULDER | A. | The Queen's Royal Regiment attached |
| 39759 | 23.01.53 | 22571611 | NEWALL | J.J. | The Queen's Royal Regiment attached |

| London Gazette issue No. | London Gazette date | Personal Service number | Name | Christian names | Remarks |
|---|---|---|---|---|---|
| THE KING'S LIVERPOOL REGIMENT | | | | | The Leather Hats raised 1685 |
| MBE: MEMBER OF THE ORDER OF THE BRITISH EMPIRE | | | | | |
| MAJOR | | | | | |
| 40036 | 04.12.53 | 380035 | HORSFORD, DSO | Derek Gordon Thomond | |
| WARRANT OFFICER CLASS I | | | | | |
| 40036 | 04.12.53 | 271770 | THOMPSON | Dennis William | |
| WARRANT OFFICER CLASS II | | | | | |
| 39870 | 01.06.53 | 3768918 | WILSON | Walter | |
| MC: MILITARY CROSS | | | | | |
| 2ND LIEUTENANTS | | | | | |
| 40036 | 04.12.53 | 424839 | DEAVILLE | Neville Frederick | |
| 40036 | 04.12.53 | 421516 | HOBSON | Patrick John Bogan | |
| 40036 | 04.12.53 | 423587 | SALMON | Alec George Broome | South Wales Borderers attached |
| 40036 | 04.12.53 | 420941 | UWINS | Anthony Brian | |
| 39906 | 07.07.53 | 418824 | WILLIAMS | John St Maur | |
| MM: MILITARY MEDAL | | | | | |
| CORPORAL | | | | | |
| 39853 | 15.05.53 | 22614193 | TURNER | Peter John | |
| PRIVATES | | | | | |
| 40036 | 04.12.53 | 22692390 | ELLISON | Edward David | Royal Inniskilling Fusiliers attached |
| 40036 | 04.12.53 | 22621410 | HARVEY | Derek | |
| MENTION IN DESPATCHES (FOR GALLANT AND DISTINGUISHED SERVICES) | | | | | |
| BRONZE OAK LEAF EMBLEM | | | | | |
| MAJOR | | | | | |
| 40036 | 04.12.53 | 255609 | QUINN, MBE | E. | |
| LIEUTENANTS | | | | | |
| 40036 | 04.12.53 | 411900 | BUCHANAN | J.R. | |
| 39930 | 14.08.53 | 412186 | CROSS | R.A. | |
| 2ND LIEUTENANTS | | | | | |
| 40036 | 04.12.53 | 426123 | DuRANDT | J. | |
| 39831 | 24.04.53 | 417251 | McBRIDE | A.J. | listed: killed in action |
| 39831 | 24.04.53 | 418329 | MacWILLIAM | W.J. | |
| 39870 | 01.06.53 | 420743 | PILE | D.A. | |

| London Gazette issue No. | London Gazette date | Personal Service number | Name | Christian names | Remarks |
|---|---|---|---|---|---|
| MENTION IN DESPATCHES (FOR GALLANT AND DISTINGUISHED SERVICES) (continued) | | | | | |
| BRONZE OAK LEAF EMBLEM | | | | | |
| WARRANT OFFICERS CLASS II | | | | | |
| 40036 | 04.12.53 | 3384850 | HARMAN | W. | East Lancashire Regiment attached |
| 40036 | 04.12.53 | 3854918 | McDORTCH | H. | The Loyal Regiment attached |
| 40036 | 04.12.53 | 3770435 | WATTLEWORTH | H.F. | |
| SERGEANTS | | | | | |
| 39870 | 01.06.53 | 19030937 | CANN | D. | |
| 39831 | 24.04.53 | 6757679 | PEARSON | A. | |
| 40036 | 04.12.53 | 22372509 | RAWLINGS | T. | |
| CORPORALS | | | | | |
| 39938 | 14.08.53 | 14457507 | ASHTON | W.A. | The Manchester Regiment attached |
| 40206 | 18.06.54 | 22539421 | GILLELAND | B.J. | listed: missing |
| PRIVATE | | | | | |
| 40036 | 04.12.53 | 22607808 | ALDRITT | J. | |

| London Gazette issue No. | London Gazette date | Personal Service number | Name | Christian names | Remarks |
|---|---|---|---|---|---|
| **THE ROYAL NORFOLK REGIMENT** | | | | | The Fighting Ninth; The Holy Boys<br>Raised 1685 |
| **DSO: DISTINGUISHED SERVICE ORDER** | | | | | |
| **MAJOR** | | | | | |
| 39831 | 24.04.53 | 291099 | CHAPMAN | Benjamin David | |
| **OBE: OFFICER OF THE ORDER OF THE BRITISH EMPIRE** | | | | | |
| **LIEUTENANT-COLONEL** | | | | | |
| 39831 | 24.04.53 | 36739 | ORLEBAR | John Hatton Roly | Officer Commanding |
| **MC: MILITARY CROSS** | | | | | |
| **CAPTAIN** | | | | | |
| 39641 | 09.09.52 | 303313 | EBERHARDIE | Charles Edward | |
| **2ND LIEUTENANTS** | | | | | |
| 39511 | 04.07.52 | 417239 | HENSON | Thomas John Brooke | |
| 39831 | 24.04.53 | 416275 | SHUTTLEWORTH | Peter Allen | |
| 39589 | 04.07.52 | 412082 | TOWELL | Anthony Philip | |
| **MM: MILITARY MEDAL** | | | | | |
| **SERGEANT** | | | | | |
| 39831 | 24.04.53 | 14731830 | BLACKBURN | Thomas | |
| **CORPORAL** | | | | | |
| 39831 | 24.04.53 | 22456064 | MAGNAY | John Frederick | |
| **PRIVATES** | | | | | |
| 39511 | 08.04.52 | 14190302 | CRITCHER | Reginald Stanley | The Essex Regiment attached |
| 39641 | 09.09.52 | 22434594 | PEARMAN | Stanley William | The Essex Regiment attached |
| 39641 | 09.09.52 | 22423697 | REED | George Raymond | |
| **MENTION IN DESPATCHES (FOR GALLANT AND DISTINGUISHED SERVICES)** | | | | | |
| **BRONZE OAK LEAF EMBLEM** | | | | | |
| **MAJORS** | | | | | |
| 39666 | 10.10.52 | 291099 | CHAPMAN | B.D. | |
| 39666 | 10.10.52 | 74611 | HOLDEN, MC | H.R. | |
| **CAPTAIN** | | | | | |
| 39641 | 09.11.52 | 370747 | McBRIDE | S.G. | |
| **LIEUTENANT** | | | | | |
| 39831 | 24.04.53 | 407758 | BERNEY | J.R.E. | listed: killed in action |
| **2ND LIEUTENANT** | | | | | |
| 39666 | 10.10.52 | 414882 | HERRING | F.E. | |

| London Gazette issue No. | London Gazette date | Personal Service number | Name | Christian names | Remarks |
|---|---|---|---|---|---|
| MENTION IN DESPATCHES (FOR GALLANT AND DISTINGUISHED SERVICES) (continued) | | | | | |
| BRONZE OAK LEAF EMBLEM | | | | | |
| WARRANT OFFICER CLASS I | | | | | |
| 39831 | 24.04.53 | 2717907 | GILCHRIST, DCM | W.J. | |
| WARRANT OFFICERS CLASS II | | | | | |
| 39666 | 10.10.52 | 14703709 | BLOOD | R.A. | |
| 39831 | 24.04.53 | 5770597 | HOWARD | C.E. | |
| | | 5827243 | WILSON | F.G.W. | The Suffolk Regiment attached |
| PRIVATES | | | | | |
| 39589 | 04.07.52 | 22405013 | NUTTER | K.E. | |
| 39502 | 28.03.52 | 22403812 | TEARLE | A.E. | |

## AWARD CONFERRED BY THE PRESIDENT OF THE UNITED STATES OF AMERICA

| | | | | | |
|---|---|---|---|---|---|
| BRONZE STAR | | | | | |
| MAJOR | | | | | |
| 39999 | 27.10.53 | 188441 | CRAMPTON | John Arthur Henry | |

| London Gazette issue No. | London Gazette date | Personal Service number | Name | Christian names | Remarks |
|---|---|---|---|---|---|
| THE ROYAL LEICESTERSHIRE REGIMENT | | | | | Bengal Tigers; Lily Whites Raised 1688 |
| DSO: DISTINGUISHED SERVICE ORDER | | | | | |
| LIEUTENANT-COLONEL | | | | | |
| 39666 | 10.10.52 | 47573 | HUTCHINS | Graham Edward Page | Officer Commanding |
| MBE: MEMBER OF THE ORDER OF THE BRITISH EMPIRE | | | | | |
| CAPTAIN | | | | | |
| 39666 | 10.10.52 | 304694 | PARSONS | Arthur John | |
| MC: MILITARY CROSS | | | | | |
| MAJOR | | | | | |
| 39420 | 28.12.51 | 66809 | DOCKER | Ludford Robert | Royal Warwickshire Regiment attached |
| LIEUTENANTS | | | | | |
| 39589 | 04.07.52 | 407771 | BREWER | Michael Herbert | |
| 39528 | 29.04.52 | 403403 | BURKE | Anthony Guyton | York & Lancaster Regiment attached |
| 2ND LIEUTENANTS | | | | | |
| 39442 | 18.01.52 | 411964 | HAVILLAND | Geoffrey Arthur | |
| 39666 | 10.10.52 | 417240 | HOLMES | Peter Fenwick | |
| MM: MILITARY MEDAL | | | | | |
| SERGEANTS | | | | | |
| 39528 | 29.04.52 | 22220718 | COX | Sidney Mervyn | listed: later died |
| 39528 | 29.04.52 | 19045177 | JENKS | Nigel Henry Peter | Sherwood Foresters attached |
| CORPORALS | | | | | |
| 39420 | 28.12.51 | 14421867 | APPLIN | Arthur George | |
| 39528 | 29.04.52 | 22276046 | MORRIS | John Thomas | Royal Warwickshire Regiment attached |
| 39589 | 04.07.52 | 19036308 | TAYLOR | Charles Alexander | |
| PRIVATES | | | | | |
| 39511 | 08.04.52 | 22441381 | CORRIGAN | William | |
| 39528 | 29.04.52 | 22229895 | WILLIAMSON | Stanley | |
| MENTION IN DESPATCHES (FOR GALLANT AND DISTINGUISHED SERVICE - POSTHUMOUS) | | | | | |
| BRONZE OAK LEAF EMBLEM | | | | | |
| SERGEANT | | | | | |
| 39528 | 29.04.52 | 14491054 | PARR | E.W. | |

| London Gazette issue No. | London Gazette date | Personal Service number | Name | Christian names | Remarks |
|---|---|---|---|---|---|
| MENTION IN DESPATCHES (FOR GALLANT AND DISTINGUISHED SERVICES) | | | | | |
| BRONZE OAK LEAF EMBLEM | | | | | |
| MAJORS | | | | | |
| 39666 | 10.10.52 | 129061 | BRIGGS, MC | B.W. | |
| 39666 | 10.10.52 | 240401 | CREAGH | J.P.N. | |
| 39666 | 10.10.52 | 67123 | MARSHALL | J.W.B. | |
| CAPTAINS | | | | | |
| 39938 | 14.08.53 | 265913 | EVANS, MC | P.S. | York & Lancaster Regiment attached |
| 39938 | 14.08.53 | 359890 | O'KANE | P.J. | Royal Berkshire Regiment attached |
| LIEUTENANT QUARTERMASTER | | | | | |
| 39666 | 10.10.52 | 375949 | NEWTON | F.W. | |
| LIEUTENANT | | | | | |
| 39328 | 07.09.51 | 372534 | COOPER | S.W. | |
| 2ND LIEUTENANT | | | | | |
| 39531 | 02.05.52 | 416587 | TOWNSHEND | R.J. | York & Lancaster Regiment attached |
| WARRANT OFFICERS CLASS II | | | | | |
| 39531 | 02.05.52 | 823559 | LAWRENCE | F.J.deM. | Sherwood Foresters attached |
| 39666 | 10.10.52 | 4856828 | MELLORS | H.G. | |
| SERGEANT | | | | | |
| 39666 | 10.10.52 | 14473580 | SEARCEY | H. | Sherwood Foresters attached |
| CORPORAL | | | | | |
| 39531 | 02.05.52 | 19040606 | KEMP | L.J. | Royal Warwickshire Regiment attached |
| LANCE-CORPORALS | | | | | |
| 39528 | 29.04.52 | 22455077 | QUINNEY | J. | Royal Warwickshire Regiment attached |
| 39528 | 29.04.52 | 19042010 | SHALER | F.W.J. | Royal Warwickshire Regiment attached |
| 39528 | 29.04.52 | 22430529 | WEBSTER | C. | Sherwood Foresters attached |
| PRIVATES | | | | | |
| 39528 | 29.04.52 | 22028406 | CROCKER | H. | listed: later died of wounds |
| 39531 | 02.05.52 | 14468529 | RANDELL | A.E. | |
| 39528 | 29.04.52 | 22471143 | TAYLOR | K. | |
| 39531 | 02.05.52 | 22447332 | TOPPING | H.S. | listed: later killed in action |

| London Gazette issue No. | London Gazette date | Personal Service number | Name | Christian names | Remarks |
|---|---|---|---|---|---|
| THE KING'S OWN SCOTTISH BORDERERS | | | | | The Brothers; Kokky-Olly-Birds Raised 1689 |
| VC: VICTORIA CROSS | | | | | |
| PRIVATE | | | | | |
| 39418 | 25.12.51 | 14471590 | SPEAKMAN | William | Black Watch (Royal Highland Regiment) attached |
| DSO: DISTINGUISHED SERVICE ORDER | | | | | |
| LIEUTENANT-COLONELS | | | | | |
| 39398 | 30.11.51 | 38523 | McDONALD, OBE | John Frederick Matheson | Officer Commanding April - October 1951 |
| 39666 | 10.10.52 | 44196 | TADMAN, OBE | Dennis Herbert | Officer Commanding October 1951 |
| MAJORS | | | | | |
| 39420 | 28.12.51 | 64605 | HARRISON | Philip Frank St Clair | |
| 39528 | 29.04.52 | 73928 | ROBERTSON-MACLEOD | Roderick Cameron | |
| 2nd LIEUTENANT | | | | | |
| 39420 | 28.12.51 | 414700 | PURVES | William | |
| MBE: MEMBER OF THE ORDER OF THE BRITISH EMPIRE | | | | | |
| MAJOR | | | | | |
| 39666 | 10.10.52 | 74640 | JACKSON, MC | Alan Edward Seaton | |
| CAPTAIN | | | | | |
| 39528 | 29.04.52 | 372744 | MILLIGAN | Kenneth William | |
| MC: MILITARY CROSS | | | | | |
| CAPTAIN | | | | | |
| 39406 | 11.12.51 | 94901 | IRVINE | Richard Henry Stewart | |
| LIEUTENANT | | | | | |
| 39666 | 10.10.52 | 407832 | FOULIS | John Alastair | |
| 2ND LIEUTENANTS | | | | | |
| 39420 | 28.12.51 | 414691 | BROOKS | Ralph Alastair | |
| 39406 | 11.12.51 | 415103 | MUDIE | Eric Robert | |
| DCM: DISTINGUISHED CONDUCT MEDAL | | | | | |
| WARRANT OFFICER CLASS II | | | | | |
| 39666 | 10.10.52 | 3189012 | MURDOCH | James | |
| SERGEANT | | | | | |
| 39420 | 28.12.51 | 14473928 | LANCASTER | John Moss | |
| CORPORAL | | | | | |
| 39398 | 30.11.51 | 22218758 | LAIDLAW | Archibald | |

| London Gazette issue No. | London Gazette date | Personal Service number | Name | Christian names | Remarks |
|---|---|---|---|---|---|
| MM: MILITARY MEDAL | | | | | |
| SERGEANTS | | | | | |
| 39328 | 07.09.51 | 2754870 | HAYE | James | Black Watch (Royal Highland Regiment) attached |
| 39420 | 28.12.51 | 22256345 | MITCHELL | Randolf Frank | Black Watch (Royal Highland Regiment) attached |
| CORPORALS | | | | | |
| 39666 | 10.10.52 | 22391717 | COMMON | Smyttan Jeffrey | |
| 39406 | 11.12.51 | 22546911 | DEVENNY | Patrick | Argyll & Sutherland Highlanders attached |
| 39442 | 18.01.52 | 22218506 | WOOD | Eric | Seaforth Highlanders attached |
| PRIVATES | | | | | |
| 39528 | 29.04.52 | 22383058 | BRYSON | Robert Ernest | |
| 39442 | 18.01.52 | 6985256 | BUCHANAN | Edward Sproule | |
| 39574 | 17.06.52 | 22454139 | CARSTAIRS | David | |
| 39328 | 07.09.51 | 22377071 | DALZIEL | David Baird | |
| 39328 | 07.09.51 | 22355707 | JOHNSTONE | Thomas | |
| 39406 | 11.12.51 | 22218180 | McCURDY | Michael | Argyll & Sutherland Highlanders attached |
| 39420 | 28.12.51 | 22202755 | PENDER | John Rodger | |
| BEM: BRITISH EMPIRE MEDAL | | | | | |
| WARRANT OFFICER CLASS II (RQMS) | | | | | |
| 39666 | 10.10.52 | 14806650 | GRANT | George Jessiman | |
| MENTION IN DESPATCHES (FOR GALLANT AND DISTINGUISHED SERVICES) | | | | | |
| BRONZE OAK LEAF EMBLEM | | | | | |
| MAJORS | | | | | |
| 39666 | 10.10.52 | 113790 | BRIGHT | K.D. | Royal Scots Fusiliers attached |
| 39398 | 30.11.51 | 375142 | LITTLE, MC | T. | Royal Scots Fusiliers attached |
| 39328 | 07.09.51 | 65532 | MACKENZIE | A.D. | Royal Scots Fusiliers attached |
| CAPTAINS | | | | | |
| 39531 | 02.05.52 | 393108 | CRAN | A.B. | |
| 39666 | 10.10.52 | 108169 | THORBURN | A.M. | |
| 39666 | 10.10.52 | 375584 | WARD | D.C.R. | |
| LIEUTENANTS | | | | | |
| 39666 | 10.10.52 | 407798 | DALRYMPLE-HAY | J.H. | Royal Scots Fusiliers attached |
| 39328 | 07.09.51 | 397314 | McMILLAN-SCOTT | A.H.F. | |
| 39528 | 29.04.52 | 408028 | ROOKE | P.B. de T. | |
| 39528 | 29.04.52 | 385974 | WILSON | K. | Royal Scots Fusiliers attached |

| London Gazette issue No. | London Gazette date | Personal Service number | Name | Christian names | Remarks |
|---|---|---|---|---|---|
| MENTION IN DESPATCHES (FOR GALLANT AND DISTINGUISHED SERVICES) (continued) | | | | | |
| BRONZE OAK LEAF EMBLEM | | | | | |
| 2ND LIEUTENANTS | | | | | |
| 39328 | 07.09.51 | 407832 | FOULIS, MC | J.A. | |
| 39328 | 07.09.51 | 409923 | INNES | C.I.K. | |
| COLOUR-SERGEANT | | | | | |
| 39666 | 10.10.52 | 14458532 | LYALL | R. | |
| SERGEANTS | | | | | |
| 39666 | 10.10.52 | 3130168 | DORWARD | J. | |
| 39398 | 30.11.51 | 14961730 | FORSTER | J.J.A. | Black Watch (Royal Highland Regiment) attached |
| 39528 | 29.04.52 | 14185662 | MUNN | A.R. | Royal Scots Fusiliers attached listed: later killed in action |
| CORPORALS | | | | | |
| 39666 | 10.10.52 | 2767263 | CASTLE | J.G. | |
| 39398 | 30.11.51 | 2233946 | McLELLAN | D.E. | |
| PRIVATES | | | | | |
| 39666 | 10.10.52 | 14466119 | BELL | T.J. | Argyll & Sutherland Highlanders attached |
| 39666 | 10.10.52 | 20401135 | KELLY | D.F. | |
| 39420 | 28.12.51 | 22406503 | McCROSSAN | G. | |
| 39528 | 24.04.52 | 22401155 | ROBERTSON | N. | |
| | | not listed | WHALLEY | D. | Royal Scots Fusiliers attached |
| 39420 | 28.12.51 | 2890080 | WILSON | S. | |

| London Gazette issue No. | London Gazette date | Personal Service number | Name | Christian names | Remarks |
|---|---|---|---|---|---|
| THE GLOUCESTERSHIRE REGIMENT | | | | | Fore and Aft; Back Numbers raised 1694 |
| VC: VICTORIA CROSS | | | | | |
| LIEUTENANT-COLONEL | | | | | |
| 39994 | 27.10.53 | 33647 | CARNE, DSO | James Power | Officer Commanding |
| LIEUTENANT | | | | | |
| 40029 | 27.11.53 | 365680 | CURTIS | Philip Kenneth Edward | Duke of Cornwall's Light Infantry attached |
| GC: GEORGE CROSS | | | | | |
| (FOR CONDUCT OF THE HIGHEST GALLANTRY WHILST A PRISONER OF WAR - POSTHUMOUS) | | | | | |
| LIEUTENANT | | | | | |
| 40146 | 13.04.54 | 463718 | WATERS | Terence Edward | West Yorkshire Regiment attached |
| DSO: DISTINGUISHED SERVICE ORDER | | | | | |
| LIEUTENANT-COLONEL | | | | | |
| 39285 | 13.07.51 | 33647 | CARNE | James Power | Officer Commanding |
| MAJOR | | | | | |
| 40036 | 04.12.53 | 67136 | HARDING | Edgar Denis | |
| CAPTAIN | | | | | |
| 40036 | 04.12.53 | 251309 | FARRAR-HOCKLEY, MC | Anthony Heritage | |
| OBE: OFFICER OF THE ORDER OF THE BRITISH EMPIRE | | | | | |
| LIEUTENANT-COLONEL | | | | | |
| 39528 | 24.04.52 | 47575 | GRIST | Digby Berkeley Angus | Officer Commanding April, 1951 |
| MBE: MEMBER OF THE ORDER OF THE BRITISH EMPIRE | | | | | |
| (FOR GALLANT AND DISTINGUISHED SERVICES WHILST A PRISONER IN NORTH KOREA) | | | | | |
| MAJOR | | | | | |
| 40146 | 09.04.54 | 126470 | WELLER | Patrick William | |
| WARRANT OFFICER CLASS I | | | | | |
| 40036 | 04.12.53 | | HOBBS | Jack Edward | |
| MC: MILITARY CROSS | | | | | |
| CAPTAINS | | | | | |
| 39273 | 29.06.51 | 360192 | HARVEY | Maurice George | The Hampshire Regiment attached |
| 39190 | 03.04.51 | 342007 | MARDELL, MM | Reginald Stanley | |
| 39528 | 29.04.51 | 393213 | MARTIN | Robert James | |

| London Gazette issue No. | London Gazette date | Personal Service number | Name | Christian names | Remarks |
|---|---|---|---|---|---|
| MC: MILITARY CROSS (continued) | | | | | |
| LIEUTENANTS | | | | | |
| 40036 | 04.12.53 | 379283 | COSTELLO | Geoffrey Tom | The Hampshire Regiment attached |
| 40036 | 04.12.53 | 400087 | TEMPLE | Guy Fredrick Bertram | |
| DCM: DISTINGUISHED CONDUCT MEDAL | | | | | |
| WARRANT OFFICER CLASS II | | | | | |
| 40036 | 04.12.53 | 5182071 | GALLAGHER | Harry | |
| SERGEANT | | | | | |
| 40036 | 04.12.53 | 5949801 | PUGH | Peter John | |
| MM: MILITARY MEDAL | | | | | |
| SERGEANTS | | | | | |
| 40036 | 04.12.53 | 21015022 | CLAYDEN | Thomas Fain | |
| 39190 | 03.04.51 | 14076869 | EAMES | Kenneth | Regiment's list shows killed in action |
| 40036 | 04.12.53 | 144497 | ROBINSON | Sydney | listed: later died |
| CORPORAL | | | | | |
| 39282 | 10.07.51 | 21125637 | WATERIDGE | Laurence Earl | |
| PRIVATES | | | | | |
| 39273 | 29.06.51 | 1430821 | CLEVELAND | Walter Stephen | The Suffolk Regiment attached |
| 40036 | 04.12.53 | 5251663 | EDWARDS | Sidney | |
| 40036 | 04.12.53 | 22530094 | MIDDLETON | Robert Leslie | |
| 40036 | 04.12.53 | 6103948 | ROBSON | James Arthur Walter | |
| 40036 | 04.12.53 | 22530161 | WALKER | Douglas Michael Robertson | |
| BEM: BRITISH EMPIRE MEDAL | | | | | |
| (FOR GALLANT AND DISTINGUISHED SERVICES WHILST A PRISONER OF WAR) | | | | | |
| CORPORALS | | | | | |
| 40146 | 09.04.54 | 22139146 | HARTIGAN | John Allen | The Hampshire Regiment attached |
| 40146 | 09.04.54 | 22530120 | HOLDHAM, MM | Alfred Edwin | |
| 40146 | 09.04.54 | 14944765 | WALTERS | Kenneth | The Hampshire Regiment attached |
| PRIVATES | | | | | |
| 40146 | 09.04.54 | 6087714 | GODDEN | Peter Leverstone | |
| 40146 | 09.04.54 | 22530240 | GODWIN | Keith Vincent | |
| 40146 | 09.04.54 | 22194057 | HAINES | David Michael | |
| 40146 | 09.04.54 | 22530171 | RICHARDS | Derek Christopher | |
| 40206 | 18.06.54 | 14468948 | STOCKTING | Donald Charles | |

| London Gazette issue No. | London Gazette date | Personal Service number | Name | Christian names | Remarks |
|---|---|---|---|---|---|
| BEM: BRITISH EMPIRE MEDAL | | | | | |
| WARRANT OFFICER CLASS II | | | | | |
| 40036 | 04.12.53 | 5182247 | MORTON | Albert Edward | |
| SERGEANTS | | | | | |
| 40036 | 04.12.53 | 14462559 | SMYTH | William James | |
| 40036 | 04.12.53 | 6203560 | SYKES | Albert | |
| MENTION IN DESPATCHES (FOR GALLANT AND DISTINGUISHED SERVICES WHILST PRISONERS OF WAR - POSTHUMOUS) | | | | | |
| BRONZE OAK LEAF EMBLEM | | | | | |
| MAJOR | | | | | |
| 40036 | 04.12.53 | 108174 | ANGIER | P.A. | |
| LIEUTENANT | | | | | |
| 40036 | 04.12.53 | 397831 | CABRAL | H.C. | |
| COLOUR-SERGEANT | | | | | |
| 40036 | 04.12.53 | 51882258 | BUXCEY | H.E. | |
| SERGEANT | | | | | |
| 40036 | 04.12.53 | 14462575 | NORTHEY | D. | |
| CORPORAL | | | | | |
| 40036 | 04.12.53 | 22288886 | HURST | R.V.L. | |
| LANCE-CORPORAL | | | | | |
| 40036 | 04.12.53 | 22530093 | BALDWIN | K.W. | |
| PRIVATES | | | | | |
| 40036 | 04.12.53 | 14405268 | BARBER | D.N. | |
| 40036 | 04.12.53 | 13021614 | BARCLAY | F.T. | |
| 40036 | 04.12.53 | 5049612 | CAIN | J. | |
| 40036 | 04.12.53 | 6290154 | DIX | W.E. | |
| 40036 | 04.12.53 | 22289532 | GALLOP | B.G. | |
| 40036 | 04.12.53 | 22249497 | MADGWICK | E.C. | |
| 40036 | 04.12.53 | 5186750 | ROBERTS | W. | |
| MENTION IN DESPATCHES (FOR GALLANT AND DISTINGUISHED SERVICES WHILST PRISONERS OF WAR) | | | | | |
| BRONZE OAK LEAF EMBLEM | | | | | |
| MAJOR | | | | | |
| 40146 | 09.04.54 | 67136 | HARDING, DSO | E.D. | |
| CAPTAIN | | | | | |
| 40146 | 09.04.54 | 251309 | FARRAR-HOCKLEY, DSO, MC | A.H. | |
| SERGEANTS | | | | | |
| 40146 | 09.04.54 | 5570676 | BOUGHTON, DCM | E. | DCM not shown in LG but is shown in "The Imjin Roll" |
| 40146 | 09.04.54 | 14489227 | HOPER | P.J. | Listed as a Private soldier in "The Imjin Roll" |

| London Gazette issue No. | London Gazette date | Personal Service number | Name | Christian names | Remarks |
|---|---|---|---|---|---|
| MENTION IN DESPATCHES (FOR GALLANT AND DISTINGUISHED SERVICES WHILST PRISONERS OF WAR) (continued) | | | | | |
| BRONZE OAK LEAF EMBLEM | | | | | |
| CORPORALS | | | | | |
| 40146 | 09.04.54 | 5619823 | BAILEY | C.A. | |
| 40146 | 09.04.54 | 5193476 | DONOHUE, MM | A.J. | |
| 40146 | 09.04.54 | 6143171 | UPJOHN | F. | |
| PRIVATES | | | | | |
| 40146 | 09.04.54 | 51101108 | FLYNN | R.G. | |
| 40206 | 18.06.54 | 19031511 | THOMAS | R.H. | |
| MENTION IN DESPATCHES (FOR GALLANT AND DISTINGUISHED SERVICES) | | | | | |
| BRONZE OAK LEAF EMBLEM | | | | | |
| LIEUTENANT-COLONEL | | | | | |
| 39382 | 07.09.51 | 477575 | GRIST | D.B.A. | Officer Commanding April 1951 |
| MAJOR | | | | | |
| 40036 | 04.12.53 | 126470 | WELLER | P.W. | |
| CAPTAINS | | | | | |
| 40036 | 04.12.53 | 362275 | LUTYENS-HUMFREY | G.D.E. | |
| 40036 | 04.12.53 | 189335 | MORRIS | W.L.D. | |
| CAPTAIN QUARTERMASTER | | | | | |
| 39328 | 07.09.51 | 185194 | WORLOCK | F.H. | |
| LIEUTENANTS | | | | | |
| 40036 | 04.12.53 | 403658 | PRESTON | A.C.N. | |
| 40036 | 04.12.53 | 463718 | WATERS | T.E. | West Yorkshire Regiment attached |
| WARRANT OFFICER CLASS II | | | | | |
| 40036 | 04.12.53 | 5194281 | RIDLINGTON | J.H. | |
| COLOUR-SERGEANT | | | | | |
| 40036 | 04.12.53 | 5182070 | PANTING | R. | |
| SERGEANTS | | | | | |
| 40036 | 04.12.53 | 893725 | BRISLAND | S.J. | |
| 40036 | 04.12.53 | 14425048 | CLAXTON | J.E. | |
| 40036 | 04.12.53 | 14189382 | DAWE | D.A. | |
| 40036 | 04.12.53 | 14466999 | MURPHY | B.J. | |
| 40036 | 04.12.53 | 5183756 | PETHERICK | P.V. | |
| 40036 | 04.12.53 | 14470887 | SMITH | B.M. | |
| CORPORAL | | | | | |
| 40036 | 04.12.53 | 22118829 | MASTERS | R.Y. | |

| London Gazette issue No. | London Gazette date | Personal Service number | Name | Christian names | Remarks |
|---|---|---|---|---|---|
| MENTION IN DESPATCHES (FOR GALLANT AND DISTINGUISHED SERVICES) (continued) | | | | | |
| BRONZE OAK LEAF EMBLEM | | | | | |
| LANCE-CORPORALS | | | | | |
| 40036 | 04.12.53 | 6460633 | CRISP | J.J. | |
| 40036 | 04.12.53 | 22530245 | PATRICK | W.A. | |
| PRIVATES | | | | | |
| 40036 | 04.12.53 | 22198639 | ALLUM | A.J. | |
| 40036 | 04.12.53 | 5951021 | DAWSON | H.H. | |
| 40036 | 04.12.53 | 14402986 | FREEMAN | G.A.A. | |
| 40036 | 04.12.53 | 22192177 | GADD | N.A.F. | |
| 40036 | 04.12.53 | 5500001 | GLARVEY | G.W. | |
| 40036 | 04.12.53 | 6400232 | HARRIS | G.H. | |
| 40036 | 04.12.53 | 22348462 | MASKELL | P.H. | |
| 40036 | 04.12.53 | 5773220 | PALFREY | W. | |
| 40036 | 04.12.53 | 75572 | PATTERSON | P. | |
| 40036 | 04.12.53 | 5885486 | PEARSON | F.E. | |
| 40036 | 04.12.53 | 22184318 | RICHARDS | F.L. | |
| 40036 | 04.12.53 | 22530104 | SAINSBURY | S. | |
| 40036 | 04.12.53 | 22189601 | WHITCHURCH | B.C. | |

## AWARD CONFERRED BY THE PRESIDENT OF THE UNITED STATES OF AMERICA

| London Gazette issue No. | London Gazette date | Personal Service number | Name | Christian names | Remarks |
|---|---|---|---|---|---|
| DISTINGUISHED SERVICE CROSS (USA) | | | | | |
| LIEUTENANT-COLONEL | | | | | |
| 39999 | 27.10.53 | 33647 | CARNE, VC, DSO | James Power | Officer Commanding |

| London Gazette issue No. | London Gazette date | Personal Service number | Name | Christian names | Remarks |
|---|---|---|---|---|---|
| THE DUKE OF WELLINGTON'S REGIMENT (WEST RIDING) | | | | | The Dukes; The Havercake Lads Raised 1702 |
| BAR TO DSO: DISTINGUISHED SERVICE ORDER | | | | | |
| LIEUTENANT-COLONEL | | | | | |
| 39906 | 07.07.53 | 44819 | BUNBURY, DSO | Francis Ramsey St Pierre | Officer Commanding |
| DSO: DISTINGUISHED SERVICE ORDER | | | | | |
| MAJOR | | | | | |
| 39906 | 07.07.53 | 66461 | KERSHAW | Lewis Francis Hardern | |
| MBE: MEMBER OF THE ORDER OF THE BRITISH EMPIRE | | | | | |
| MAJOR | | | | | |
| 39870 | 01.06.53 | 62616 | SKELSEY | Walter | |
| WARRANT OFFICER CLASS II | | | | | |
| 40036 | 04.12.53 | 4611264 | HALL | John Stanley | |
| BAR TO MC: MILITARY CROSS | | | | | |
| MAJOR | | | | | |
| 40036 | 04.12.53 | 95620 | KAVANAGH, MC | Aveling Barry Martin | |
| MC: MILITARY CROSS | | | | | |
| MAJORS | | | | | |
| 40036 | 04.12.53 | 62586 | AUSTIN | Rudolph Edmund | |
| 39906 | 07.07.53 | 151631 | EMETT | Edward Joseph Parr | |
| 40036 | 04.12.53 | 85464 | FIRTH, MBE | Anthony Denys | |
| LIEUTENANTS | | | | | |
| 40036 | 04.12.53 | 409760 | BORWELL | David Leslie | |
| 40036 | 04.12.53 | 419724 | GILBERT-SMITH | David Stuart | |
| 39788 | 27.02.53 | 407858 | HARMS | Rodney Malcolm | |
| 40036 | 04.12.53 | 423800 | HOLLANDS | Douglas John | |
| 2ND LIEUTENANTS | | | | | |
| 39788 | 27.02.53 | 423582 | ORR | Patrick Ian | |
| 40036 | 04.12.53 | 418393 | STACPOOLE | Humphrey Adam John | |
| MM: MILITARY MEDAL | | | | | |
| WARRANT OFFICER CLASS II | | | | | |
| 39906 | 07.07.53 | 4388113 | JOBLING | Joseph Charles | The Green Howards attached |

| London Gazette issue No. | London Gazette date | Personal Service number | Name | Christian names | Remarks |
|---|---|---|---|---|---|
| MM: MILITARY MEDAL (continued) | | | | | |
| SERGEANT | | | | | |
| 40036 | 04.12.53 | 22249555 | NOWELL | Thomas | York & Lancaster Regiment attached |
| CORPORALS | | | | | |
| 40036 | 04.12.53 | 22795542 | McKENZIE | Henry Anthony | The Green Howards attached |
| 39906 | 07.07.53 | 22525040 | PICKERSGILL | George | The East Yorkshire Regiment attached |
| 39906 | 07.07.53 | 22585248 | WALKER | John | |
| LANCE-CORPORAL | | | | | |
| 39898 | 26.06.53 | 22550987 | BAILEY | Herbert | |
| PRIVATE | | | | | |
| 39906 | 07.07.53 | 22618200 | HUSBAND | Denis Wilfred | |
| MENTION IN DESPATCHES (FOR GALLANT AND DISTINGUISHED SERVICES - POSTHUMOUS) | | | | | |
| BRONZE OAK LEAF EMBLEM | | | | | |
| PRIVATE | | | | | |
| 39938 | 14.08.53 | 22596217 | CONNOR | M. | |
| MENTION IN DESPATCHES (FOR GALLANT AND DISTINGUISHED SERVICES) | | | | | |
| BRONZE OAK LEAF EMBLEM | | | | | |
| MAJORS | | | | | |
| 40036 | 04.12.53 | 85713 | INCE | R.H. | |
| 40036 | 04.12.53 | 112871 | LE MESSURIER | H.S. | |
| 40036 | 04.12.53 | 50937 | MORAN, OBE | R. dela H. | |
| CAPTAIN | | | | | |
| 40036 | 04.12.53 | 362946 | ROBERTSON | W.F.C. | |
| LIEUTENANTS | | | | | |
| 40036 | 04.12.53 | 419011 | GLEN | C.H. | listed: later died |
| 40036 | 04.12.53 | 393157 | HARDY | E.M.P. | |
| 40036 | 04.12.53 | 414929 | NAUGHTON | J.N.H. | |
| WARRANT OFFICER CLASS I | | | | | |
| 39870 | 01.06.53 | 4743875 | PEARCE | R. | |
| WARRANT OFFICER CLASS II | | | | | |
| 40036 | 04.12.53 | 15001316 | CORKE | G. | |
| SERGEANT | | | | | |
| 40036 | 04.12.53 | 10939172 | ROBINS | W. | |

| London Gazette issue No. | London Gazette date | Personal Service number | Name | Christian names | Remarks |
|---|---|---|---|---|---|
| MENTION IN DESPATCHES (FOR GALLANT AND DISTINGUISHED SERVICES) (continued) | | | | | |
| BRONZE OAK LEAF EMBLEM | | | | | |
| LANCE-CORPORAL | | | | | |
| 39938 | 14.08.53 | 22590589 | SMITH | D.C. | |
| PRIVATES | | | | | |
| 39938 | 14.08.53 | 22636205 | DAVIES | E.J. | |
| 39938 | 14.08.53 | 22596845 | JOHNSTONE | G.S. | |
| 39938 | 14.08.53 | 22610617 | QUARNBY | L. | also listed: The Welch Regiment |

## AWARD CONFERRED BY THE PRESIDENT OF THE UNITED STATES OF AMERICA

| London Gazette issue No. | London Gazette date | Personal Service number | Name | Christian names | Remarks |
|---|---|---|---|---|---|
| BRONZE STAR | | | | | |
| MAJOR | | | | | |
| 40249 | 06.08.54 | 85713 | INCE | Richard Henry | |

| London Gazette issue No. | London Gazette date | Personal Service number | Name | Christian names | Remarks |
|---|---|---|---|---|---|
| THE WELCH REGIMENT | | | | | 41st First Invalids<br>69th Ups and Downs<br>Raised 1719 |
| DSO: DISTINGUISHED SERVICE ORDER | | | | | |
| LIEUTENANT-COLONEL | | | | | |
| 39831 | 24.04.53 | 39437 | DEANE | Henry Hargrave | Officer Commanding |
| MAJORS | | | | | |
| 39634 | 29.08.52 | 68091 | JACKSON | Thomas John | |
| 39831 | 24.04.53 | 74666 | ROBERTS | Arthur Gordon | |
| MC: MILITARY CROSS | | | | | |
| 2ND LIEUTENANTS | | | | | |
| 39589 | 04.07.52 | 417045 | BOWLER | John Leslie | |
| 39831 | 24.04.53 | 417047 | DAVEY | Keith John | |
| 39666 | 10.10.52 | 414949 | POWYS | Ian Atherton | |
| MM: MILITARY MEDAL | | | | | |
| CORPORALS | | | | | |
| 39831 | 24.04.53 | 14477311 | BEER | Patrick William | The Devonshire Regiment attached |
| 39666 | 10.10.52 | 19061614 | HUGHES | David George | The South Wales Borderers attached |
| 39511 | 08.04.52 | 22446901 | WHITTLE | William Harry | |
| MENTION IN DESPATCHES (FOR GALLANT AND DISTINGUISHED SERVICE - POSTHUMOUS) | | | | | |
| BRONZE OAK LEAF EMBLEM | | | | | |
| PRIVATE | | | | | |
| 39938 | 14.08.53 | 22596217 | CONNOR | M. | |
| MENTION IN DESPATCHES (FOR GALLANT AND DISTINGUISHED SERVICES) | | | | | |
| BRONZE OAK LEAF EMBLEM | | | | | |
| CAPTAIN | | | | | |
| 39666 | 10.10.52 | 184164 | TAVERNER | R.C. | |
| WARRANT OFFICER CLASS II | | | | | |
| 39666 | 10.10.52 | 3957997 | CANNON | S.J. | |
| SERGEANTS | | | | | |
| 39589 | 04.07.52 | 19035713 | BLOWING | D. | The Royal Welch Fusiliers attached |
| 39831 | 24.04.53 | 3976771 | LEAHY | T. | |
| CORPORALS | | | | | |
| 39666 | 10.10.52 | 21125061 | HANNAM | E.J. | The South Wales Borderers attached |
| 39831 | 24.04.53 | 22454641 | SMITH | W.R. | |

| London Gazette issue No. | London Gazette date | Personal Service number | Name | Christian names | Remarks |
|---|---|---|---|---|---|
| MENTION IN DESPATCHES (FOR GALLANT AND DISTINGUISHED SERVICES) (continued) | | | | | |
| LANCE-CORPORALS | | | | | |
| 39666 | 10.10.52 | 22454638 | ROBERTS | A. | The Royal Welch Fusiliers attached |
| 39938 | 14.08.53 | 22590589 | SMITH | D.C. | |
| PRIVATES | | | | | |
| 39666 | 10.10.52 | 22436273 | BARNES | R.D. | East Lancashire Regiment attached |
| 39938 | 14.08.53 | 22636205 | DAVIES | E.J. | |
| 39666 | 10.10.52 | 22448948 | GRIFFITHS | E.M. | |
| 39938 | 14.08.53 | 22596845 | JOHNSTONE | G.S. | |
| 39589 | 04.07.52 | 22475024 | LEWIS | P. | |
| 39666 | 10.10.52 | 19037498 | MARSHALL | R.I. | The Gloucestershire Regiment attached |
| 40036 | 04.12.53 | 22614218 | MICHAEL | P.C. | |
| 39938 | 14.08.53 | 22610617 | QUARNBY | L. | also listed: Duke of Wellington's Regiment |
| 39831 | 24.04.53 | 22475003 | SHORT | P.A. | |

## AWARD CONFERRED BY THE PRESIDENT OF THE UNITED STATES OF AMERICA

| London Gazette issue No. | London Gazette date | Personal Service number | Name | Christian names | Remarks |
|---|---|---|---|---|---|
| BRONZE STAR | | | | | |
| MAJOR | | | | | |
| 39999 | 27.10.53 | 44883 | HALLOWES | Frederick Garnegy | |

| London Gazette issue No. | London Gazette date | Personal Service number | Name | Christian names | Remarks |
|---|---|---|---|---|---|
| THE BLACK WATCH (ROYAL HIGHLAND REGIMENT) | | | | | Forty Twas; The Watch Raised 1739 |
| BAR TO DSO: DISTINGUISHED SERVICE ORDER | | | | | |
| LIEUTENANT-COLONEL | | | | | |
| 59870 | 01.06.53 | 50948 | ROSE, DSO | David MacNeil Campbell | Officer Commanding |
| DSO: DISTINGUISHED SERVICE ORDER | | | | | |
| MAJOR | | | | | |
| 39479 | 09.01.53 | 73151 | IRWIN, MC | Angus Digby Hasting | |
| MBE: MEMBER OF THE ORDER OF THE BRITISH EMPIRE | | | | | |
| MAJOR | | | | | |
| 40036 | 04.12.53 | 66713 | MACRAE | Robert Andrew Alexander Scarth | Seaforth Highlanders attached |
| CAPTAIN QUARTERMASTER | | | | | |
| 39870 | 01.06.53 | 250692 | CLARK | Harold McLoy | |
| WARRANT OFFICER CLASS I | | | | | |
| 40036 | 04.12.53 | 2754502 | SCOTT | William | |
| MC: MILITARY CROSS | | | | | |
| MAJORS | | | | | |
| 40036 | 04.12.53 | 176941 | DENNISTON | Robert Keith | Seaforth Highlanders attached |
| 40036 | 04.12.53 | 177766 | LITHGOW | Anthony Onslow Laurence | |
| LIEUTENANT | | | | | |
| 39749 | 09.01.53 | 369889 | HAW | Richard John | |
| 2ND LIEUTENANTS | | | | | |
| 39749 | 09.01.53 | 419717 | BLACK | Michael Donald Gordon | |
| 39831 | 24.04.53 | 421594 | RATTRAY | Alexander Campbell | |
| DCM: DISTINGUISHED CONDUCT MEDAL | | | | | |
| SERGEANT | | | | | |
| 39749 | 09.01.53 | 22321842 | GAIT | Brian | |
| MM: MILITARY MEDAL | | | | | |
| SERGEANTS | | | | | |
| 39831 | 24.04.53 | 14465633 | FLOYD | Frederick George | |
| 39749 | 09.01.53 | 19036797 | HUTCHISON | Alexander Brown | |
| 39749 | 09.01.53 | 19033346 | KERRY | William Edward | |

| London Gazette issue No. | London Gazette date | Personal Service number | Name | Christian names | Remarks |
|---|---|---|---|---|---|
| MM: MILITARY MEDAL (continued) | | | | | |
| CORPORAL | | | | | |
| 40036 | 04.12.53 | 22526706 | BEATTIE | Robert William | |
| LANCE-CORPORAL | | | | | |
| 39749 | 09.01.53 | 22522253 | MANNING | Robert Alfred | |
| MENTION IN DESPATCHES (FOR GALLANT AND DISTINGUISHED SERVICES) | | | | | |
| BRONZE OAK LEAF EMBLEM | | | | | |
| MAJORS | | | | | |
| 40036 | 04.12.53 | 70611 | BUCHANAN, MBE, TD | P.G. | |
| 39870 | 01.06.53 | 105763 | MOIR | C.MacB | |
| 40036 | 04.12.53 | 105763 | MOIR | C.MacB | |
| CAPTAIN | | | | | |
| 40036 | 04.12.53 | 3377264 | NICOLL | E.W. | |
| LIEUTENANT | | | | | |
| 40036 | 04.12.53 | 403623 | MONCRIEFF | J.G. | |
| 2ND LIEUTENANTS | | | | | |
| 40036 | 04.12.53 | 423802 | CLAZY | M.G. | |
| 39831 | 24.04.53 | 417059 | NICOLL | D.A.S. | listed: killed in action |
| WARRANT OFFICER CLASS II | | | | | |
| 39870 | 01.06.53 | 2695724 | PATERSON | G. | |
| SERGEANT | | | | | |
| 40036 | 04.12.53 | 22577734 | EDMISTON | A. | |
| CORPORALS | | | | | |
| 39749 | 09.01.53 | 2755099 | STRACHAN | F. | |
| 39749 | 09.01.53 | 22200974 | McKAY | J. | Seaforth Highlanders attached |
| PRIVATES | | | | | |
| 39749 | 09.01.53 | 22509651 | COLEY | G. | Royal Leicestershire Regiment attached |
| 39749 | 09.01.53 | 22462521 | WALKER | J.N. | |

| London Gazette issue No. | London Gazette date | Personal Service number | Name | Christian names | Remarks |
|---|---|---|---|---|---|
| THE KING'S SHROPSHIRE LIGHT INFANTRY | | | | | 53rd Old Five and Three Pennies Raised 1755 |
| BAR TO DSO: DISTINGUISHED SERVICE ORDER | | | | | |
| LIEUTENANT-COLONEL | | | | | |
| 39666 | 10.10.52 | 44033 | BARLOW, DSO, OBE | Vernon William | Officer Commanding |
| DSO: DISTINGUISHED SERVICE ORDER | | | | | |
| MAJOR | | | | | |
| 39398 | 30.11.51 | 69884 | COTTLE | Wallace James | |
| MBE: MEMBER OF THE ORDER OF THE BRITISH EMPIRE | | | | | |
| MAJOR | | | | | |
| 39666 | 10.10.52 | 86405 | RADCLIFFE | Guy Lushington Yonge | |
| CAPTAINS | | | | | |
| 39666 | 10.10.52 | 172350 | GARNETT | Robert Hugh | |
| 39666 | 10.10.52 | 364093 | ROBERTS | John Robert | |
| MC: MILITARY CROSS | | | | | |
| CAPTAINS | | | | | |
| 39406 | 11.12.51 | 397804 | BALLENDEN | John Patrick St Clair | |
| 39528 | 29.04.52 | 369467 | GRUNDY | Christopher Barnes | |
| LIEUTENANTS | | | | | |
| 39406 | 11.12.51 | 397815 | BORWICK | Andrew | |
| 39574 | 17.06.51 | 411225 | PACK | Anthony George | |
| 2ND LIEUTENANTS | | | | | |
| 39442 | 18.02.52 | 412603 | BLYTH | Philip Henry | Oxfordshire and Buckinghamshire Light Infantry attached |
| 39666 | 10.10.52 | 413405 | WHYBROW | John | |
| DCM: DISTINGUISHED CONDUCT MEDAL | | | | | |
| CORPORAL | | | | | |
| 39398 | 30.11.51 | 19034149 | WADE | Leonard | |
| MM: MILITARY MEDAL | | | | | |
| WARRANT OFFICER CLASS II | | | | | |
| 39528 | 29.04.52 | 4032288 | PEALL | John Richard Thomas | |

| London Gazette issue No. | London Gazette date | Personal Service number | Name | Christian names | Remarks |
|---|---|---|---|---|---|
| MM: MILITARY MEDAL (continued) | | | | | |
| SERGEANT | | | | | |
| 39328 | 07.09.51 | 21015099 | RAISON | Peter | |
| CORPORALS | | | | | |
| 39661 | 03.10.52 | 22523164 | NEWTON | Gerald | |
| 39406 | 11.12.51 | 14731486 | PENDLEBURY | Robert | |
| 39666 | 10.10.52 | 19043240 | TALBOT | Alfred | |
| 39406 | 11.12.51 | 5383934 | WHITMORE | Clement Arthur | |
| LANCE-CORPORALS | | | | | |
| 39528 | 29.04.52 | 22366040 | FUNNELL | Peter Bernard | |
| 39406 | 11.12.51 | 22401907 | NORTON | John | |
| MENTION IN DESPATCHES (FOR GALLANT AND DISTINGUISHED SERVICES) | | | | | |
| BRONZE OAK LEAF EMBLEM | | | | | |
| LIEUTENANT-COLONEL | | | | | |
| 39528 | 29.04.52 | 44033 | BARLOW, DSO, OBE | V.W. | Officer Commanding |
| MAJORS | | | | | |
| 39661 | 03.10.52 | 69171 | BANCROFT | D.R.J. | |
| 39666 | 10.10.52 | 325909 | PHILLIPS | P.T.D. | |
| CAPTAINS | | | | | |
| 39531 | 02.05.52 | 393099 | CHAMBERS | P.M. | |
| 39398 | 30.11.51 | 372861 | HOUGHTON-BERRY | B.V. | |
| LIEUTENANT | | | | | |
| 39528 | 29.04.52 | 387907 | CLEAVER | K.A.R. | |
| WARRANT OFFICER CLASS I | | | | | |
| 39666 | 10.10.52 | 2733291 | KNIGHT, MBE | H.E. | |
| WARRANT OFFICER CLASS II | | | | | |
| 39528 | 29.04.52 | 4031509 | OLDEN, MM | W.W.F. | |
| SERGEANTS | | | | | |
| 39666 | 10.10.52 | 4693833 | FITZGERALD | M.J. | |
| 39666 | 10.10.52 | 22222912 | LEWENDON | A. | Oxfordshire and Buckinghamshire Light Infantry attached |
| 39666 | 10.10.52 | 22009430 | STIRMAN | A. | |
| CORPORALS | | | | | |
| 39666 | 10.10.52 | 19039012 | CARR | J.T.R. | Durham Light Infantry attached |
| 39398 | 30.11.51 | 14445551 | PROCTOR | D.W. | Oxfordshire and Buckinghamshire Light Infantry attached |
| 39528 | 29.04.52 | 14187165 | SLATER | C.R. | Duke of Cornwall's Light Infantry attached |

| London Gazette issue No. | London Gazette date | Personal Service number | Name | Christian names | Remarks |
|---|---|---|---|---|---|
| MENTION IN DESPATCHES (FOR GALLANT AND DISTINGUISHED SERVICES) (continued) | | | | | |
| BRONZE OAK LEAF EMBLEM | | | | | |
| LANCE-CORPORAL | | | | | |
| 39531 | 02.05.52 | 22231589 | CUTHBERT | E.E. | Durham Light Infantry attached |
| PRIVATE | | | | | |
| 39398 | 30.11.51 | 22276462 | WELLS | T.W. | Bedfordshire and Hertfordshire Regiment attached |

| London Gazette issue No. | London Gazette date | Personal Service number | Name | Christian names | Remarks |
|---|---|---|---|---|---|
| THE MIDDLESEX REGIMENT (DUKE OF CAMBRIDGE'S OWN) | | | | | Die Hards - Albuhera Raised 1755 |
| DSO: DISTINGUISHED SERVICE ORDER | | | | | |
| LIEUTENANT-COLONEL | | | | | |
| 39090 | 12.12.50 | 44909 | MAN, OBE | Andrew Morrice | Officer Commanding |
| MC: MILITARY CROSS | | | | | |
| 2ND LIEUTENANTS | | | | | |
| 39150 | 16.02.51 | 407942 | LAWRENCE | Christopher Lansdowne | |
| 39328 | 07.09.51 | 411732 | REED | Barry St George A. | |
| MM: MILITARY MEDAL | | | | | |
| PRIVATE | | | | | |
| 39207 | 20.04.51 | 2549004 | COBBY | Leslie Raymond | |
| MENTION IN DESPATCHES (FOR GALLANT AND DISTINGUISHED SERVICES) | | | | | |
| BRONZE OAK LEAF EMBLEM | | | | | |
| MAJOR | | | | | |
| 40036 | 04.12.53 | 255225 | POND, MC | H.C. | |
| CAPTAIN | | | | | |
| 39328 | 07.09.51 | 343118 | CARTER | K.J. | |
| LIEUTENANT | | | | | |
| 39164 | 06.03.51 | 370959 | SANDER | G. | listed: killed in action |
| SERGEANT | | | | | |
| 39164 | 06.03.51 | 14187910 | BERMINGHAM | E.J. | |
| CORPORAL | | | | | |
| 39164 | 06.03.51 | 6202181 | FIELD | J. | |
| | | not listed | PENTONY | J. | extracted from: "The Die-Hards in Korea" |
| LANCE-CORPORAL | | | | | |
| 39164 | 06.03.51 | 14472830 | MEAD | L.A. | |
| PRIVATES | | | | | |
| 39328 | 07.09.51 | 22157342 | LOFTS | J.A. | |
| 39164 | 06.03.51 | 14475345 | MATTHEWS | G.E.F. | |

| London Gazette issue No. | London Gazette date | Personal Service number | Name | Christian names | Remarks |
|---|---|---|---|---|---|

AWARDS CONFERRED BY THE PRESIDENT OF THE UNITED STATES OF AMERICA

LEGION OF MERIT, DEGREE OF LEGIONNAIRE

COLONEL

| | | | | | |
|---|---|---|---|---|---|
| 39999 | 27.10.53 | 44909 | MAN, DSO, OBE | Andrew Morrice | |

BRONZE STAR

MAJOR

| | | | | | |
|---|---|---|---|---|---|
| 39254 | 08.06.51 | 49853 | GWYN | Rhys Anthony | |

SERGEANT

| | | | | | |
|---|---|---|---|---|---|
| 39254 | 08.06.51 | 14490716 | HUMMERSTONE | Douglas Bryant | listed: later killed in action |

| London Gazette issue No. | London Gazette date | Personal Service number | Name | Christian names | Remarks |
|---|---|---|---|---|---|
| THE DURHAM LIGHT INFANTRY | | | | | The Faithful Durhams Raised 1758 |
| BAR TO THE DSO: DISTINGUISHED SERVICE ORDER | | | | | |
| LIEUTENANT-COLONEL | | | | | |
| 40036 | 04.12.53 | 44141 | JEFFREYS, DSO, OBE | Peter John | Officer Commanding |
| OBE: OFFICER OF THE ORDER OF THE BRITISH EMPIRE | | | | | |
| MAJOR | | | | | |
| 40036 | 04.12.53 | 36327 | ATKINSON, MC | Reginald Gordon | |
| MBE: MEMBER OF THE ORDER OF THE BRITISH EMPIRE | | | | | |
| CAPTAIN | | | | | |
| 40036 | 04.12.53 | 381263 | TONKINSON | John Bryan | |
| WARRANT OFFICERS CLASS II | | | | | |
| 40036 | 04.12.53 | 2567620 | CALVERT | George Alfred | |
| 40036 | 04.12.53 | 19673511 | GIBBENS | Charles Harold Joseph | |
| MC: MILITARY CROSS | | | | | |
| MAJOR | | | | | |
| 40036 | 04.12.53 | 326145 | SCOTT | Robert Edmund | |
| 2ND LIEUTENANTS | | | | | |
| 39938 | 14.08.53 | 421955 | CUNNINGHAM | John Charles Howard | King's Own Yorkshire Light Infantry attached |
| 40036 | 04.12.53 | 421953 | McGREGOR-OAKFORD | Robert Brian | |
| 39870 | 01.06.53 | 421969 | PEARCE | John George | King's Own Shropshire Light Infantry attached |
| MM: MILITARY MEDAL | | | | | |
| CORPORAL | | | | | |
| 40036 | 04.12.53 | 22524254 | LOFTHOUSE | Robert | King's Own Yorkshire Light Infantry attached |
| LANCE-CORPORALS | | | | | |
| 39749 | 09.01.53 | 22457807 | MOORE | Ronald Alfred | |
| 40036 | 04.12.53 | 22773552 | STOCKTON | Robert James | King's Own Yorkshire Light Infantry attached |
| PRIVATE | | | | | |
| 39938 | 14.08.53 | 22646443 | RAWLINGS | Derry Irving | |

| London Gazette issue No. | London Gazette date | Personal Service number | Name | Christian names | Remarks |
|---|---|---|---|---|---|
| BEM: BRITISH EMPIRE MEDAL | | | | | |
| SERGEANT | | | | | |
| 40036 | 04.12.53 | 22259436 | COXON | John Robert | |
| MENTION IN DESPATCHES (FOR GALLANT AND DISTINGUISHED SERVICES) | | | | | |
| BRONZE OAK LEAF EMBLEM | | | | | |
| MAJOR | | | | | |
| 39870 | 01.06.53 | 326145 | SCOTT | R.E. | |
| CAPTAIN | | | | | |
| 39831 | 24.04.53 | 360554 | BURINI | F.B. | |
| LIEUTENANT | | | | | |
| 40036 | 04.12.53 | 403673 | NOTT-BOWER | W.J. | |
| 2ND LIEUTENANTS | | | | | |
| 39831 | 24.04.53 | 419482 | PERROTT | B.D. | |
| 39831 | 24.04.53 | 420627 | SAWBRIDGE | E.H.F. | |
| WARRANT OFFICERS CLASS II | | | | | |
| 39831 | 24.04.53 | 4456113 | BEIRNE | L. | |
| 40036 | 04.12.53 | 5619295 | ELLIOTT | F.A. | King's Own Shropshire Light Infantry attached |
| CORPORALS | | | | | |
| 40036 | 04.12.53 | 22620997 | ALLEN | T.G. | |
| 39870 | 01.06.53 | 22476892 | ASPINSHAW | H.D. | |
| 40036 | 04.12.53 | 22618571 | MURRAY | J.G. | |
| 40036 | 04.12.53 | 22478698 | SHANNON | R. | |
| 40036 | 04.12.53 | 22276387 | STOKES | L. | |
| PRIVATES | | | | | |
| 39831 | 24.04.53 | 22615132 | BALL | A.H. | |
| 39938 | 14.08.53 | 22597212 | ROSEVEAR | D.A.H. | |

| London Gazette issue No. | London Gazette date | Personal Service number | Name | Christian names | Remarks |
|---|---|---|---|---|---|
| THE ROYAL ULSTER RIFLES | | | | | Fitch's Grenadiers; Irish Giants raised 1793 |
| DSO: DISTINGUISHED SERVICE ORDER | | | | | |
| MAJORS | | | | | |
| 39282 | 10.07.51 | 149563 | GAFFIKIN | Hugh Montgomery | |
| 39205 | 17.04.51 | 69180 | SHAW, MC | John Kirkpatrick Hay | |
| OBE: OFFICER OF THE ORDER OF THE BRITISH EMPIRE | | | | | |
| LIEUTENANT-COLONEL | | | | | |
| 39528 | 29.04.52 | 44059 | CARSON | Robert John Heyworth | Officer Commanding |
| MBE: MEMBER OF THE ORDER OF THE BRITISH EMPIRE (FOR GALLANT AND DISTINGUISHED SERVICES WHILST A PRISONER OF WAR IN NORTH KOREA) | | | | | |
| MAJOR | | | | | |
| 40146 | 09.03.54 | 64645 | RYAN | Michael Dennis George Conybeare | |
| MBE: MEMBER OF THE ORDER OF THE BRITISH EMPIRE | | | | | |
| MAJOR | | | | | |
| 39666 | 23.10.52 | 148939 | ANDERSON, MC | William Ellery | |
| BAR TO MC: MILITARY CROSS | | | | | |
| CAPTAIN | | | | | |
| 39666 | 23.10.52 | 148939 | ANDERSON, MC | William Ellery | |
| MC: MILITARY CROSS | | | | | |
| LIEUTENANTS | | | | | |
| 39273 | 29.06.51 | 385955 | MacNICHOL | Colin | The Border Regiment attached |
| 39282 | 10.07.51 | 402905 | TREVOR-ROPER | Anthony Dacre | The Lancashire Fusiliers attached |
| 2ND LIEUTENANTS | | | | | |
| 39282 | 10.07.51 | 407952 | McCORD | Mervyn Noel Samuel | |
| 39205 | 17.04.51 | 406701 | SHAW-STEWART | Houston Mark | |
| DCM: DISTINGUISHED CONDUCT MEDAL | | | | | |
| WARRANT OFFICER CLASS II | | | | | |
| 39288 | 17.08.51 | 7010702 | McCONVILLE | Andrew | |

| London Gazette issue No. | London Gazette date | Personal Service number | Name | Christian names | Remarks |
|---|---|---|---|---|---|
| MM: MILITARY MEDAL | | | | | |
| SERGEANTS | | | | | |
| 39205 | 17.04.51 | 22243358 | CAMPBELL | Henry Adams | |
| 39273 | 29.06.51 | 7011513 | COOKE | Daniel | |
| CORPORAL | | | | | |
| 39398 | 30.11.51 | 22229678 | HALLIDAY | William McWilliam | |
| RIFLEMAN | | | | | |
| 39398 | 30.11.51 | 22337903 | NEWPORT | John Frederick | |
| BEM: BRITISH EMPIRE MEDAL (FOR GALLANT AND DISTINGUISHED SERVICE WHILST A PRISONER OF WAR) | | | | | |
| CORPORAL | | | | | |
| 40146 | 13.04.54 | 14456352 | O'HARA | William Patrick Michael | |
| MENTION IN DESPATCHES (FOR GALLANT AND DISTINGUISHED SERVICES WHILST PRISONERS OF WAR) | | | | | |
| BRONZE OAK LEAF EMBLEM | | | | | |
| CAPTAIN | | | | | |
| 40146 | 09.04.54 | 193883 | MAJURY | J.H.S. | |
| LANCE-CORPORAL | | | | | |
| 40146 | 09.04.54 | 22202796 | MASSEY | W. | |
| RIFLEMEN | | | | | |
| 40146 | 09.04.54 | 22308421 | BUCKLEY | J.J. | |
| MENTION IN DESPATCHES (FOR GALLANT AND DISTINGUISHED SERVICES) | | | | | |
| BRONZE OAK LEAF EMBLEM | | | | | |
| LIEUTENANT-COLONEL | | | | | |
| 39328 | 07.09.51 | 44059 | CARSON | R.J.H. | Officer Commanding |
| MAJOR | | | | | |
| 39528 | 07.09.51 | 62664 | RICKCORD | G.P. | |
| CAPTAINS | | | | | |
| 39282 | 10.07.52 | 314806 | HINDE | J.E.D'O. | |
| 39666 | 10.10.52 | 273669 | JOHNSON | V.de P.C. | |
| 39273 | 29.06.51 | 200407 | NIXON, Bt., MC | Sir C.J. | |
| 39528 | 29.04.52 | 251823 | SINCLAIR | R.H.S. | |
| 39328 | 07.09.51 | 333558 | | | |
| CAPTAIN QUARTERMASTER | | | | | |
| 39328 | 07.09.51 | 333558 | SMITH | T.P. | |
| LIEUTENANTS | | | | | |
| 39398 | 30.11.51 | 373842 | CORNICK | L. | |
| 39282 | 10.07.52 | 397992 | MOLE | J.J. | |
| 39398 | 30.11.51 | 413609 | POTTS | G.L. | |

| London Gazette issue No. | London Gazette date | Personal Service number | Name | Christian names | Remarks |
|---|---|---|---|---|---|
| MENTION IN DESPATCHES (FOR GALLANT AND DISTINGUISHED SERVICES) (continued) | | | | | |
| BRONZE OAK LEAF EMBLEM | | | | | |
| COLOUR-SERGEANT | | | | | |
| 39328 | 07.09.51 | 7012597 | BYRNE | S. | |
| SERGEANTS | | | | | |
| 39328 | 07.09.51 | 14190364 | FITZSIMONS | A. | |
| 39205 | 17.04.51 | 7047229 | FOWLER | T.J. | |
| CORPORALS | | | | | |
| 39205 | 17.04.51 | 4127146 | HUNT | N. | |
| 39205 | 17.04.51 | 3654481 | WATKINSON | J. | |
| RIFLEMEN | | | | | |
| 39282 | 10.07.52 | 22511803 | McSHANE | J.J. | |
| 39528 | 29.04.52 | 6980391 | PLEWS | E. | Royal Inniskilling Fusiliers attached |
| 39205 | 17.04.51 | 6837575 | VARLEY | A.E. | |

## AWARDS CONFERRED BY THE PRESIDENT OF THE UNITED STATES OF AMERICA

| | | | | | |
|---|---|---|---|---|---|
| SILVER STAR | | | | | |
| MAJOR | | | | | |
| 39462 | 08.02.52 | 62664 | RICKCORD, DSO | Gerald Percival | |
| BRONZE STAR | | | | | |
| LIEUTENANT-COLONEL | | | | | |
| 39999 | 27.10.53 | 90469 | LAISTER, TD | Thomas Leslie | |

| London Gazette issue No. | London Gazette date | Personal Service number | Name | Christian names | Remarks |
|---|---|---|---|---|---|
| THE ARGYLL AND SUTHERLAND HIGHLANDERS (PRINCESS LOUISE'S) | | | | | The Rorys; The Thin Red Line<br>Raised 1794 |
| VC: VICTORIA CROSS - POSTHUMOUS | | | | | |
| MAJOR | | | | | |
| 39115 | 05.01.51 | 50980 | MUIR | Kenneth | Killed in action |
| DSO: DISTINGUISHED SERVICE ORDER | | | | | |
| LIEUTENANT-COLONEL | | | | | |
| 39090 | 12.12.50 | 36735 | NEILSON | George Leslie | Officer Commanding |
| OBE: OFFICER OF THE ORDER OF THE BRITISH EMPIRE | | | | | |
| MAJORS | | | | | |
| 39328 | 07.09.51 | 137940 | SLOANE | John Bromley Malet | |
| 39328 | 07.09.51 | 73991 | STEWART, MBE, TD | James Douglas | |
| MBE: MEMBER OF THE ORDER OF THE BRITISH EMPIRE | | | | | |
| CAPTAIN QUARTERMASTER | | | | | |
| 39328 | 07.09.51 | 317386 | BROWN | Andrew Wilkie | |
| BAR TO MC: MILITARY CROSS | | | | | |
| MAJOR | | | | | |
| 39084 | 05.12.50 | 88487 | PENMAN, MC, TD | John Albert | |
| MM: MILITARY MEDAL | | | | | |
| WARRANT OFFICER CLASS II | | | | | |
| 39150 | 16.02.51 | 3855700 | COLLETT | Thomas | |
| CORPORAL | | | | | |
| 39150 | 16.02.51 | 19031261 | SWEENEY | Robert Rodden | |
| MENTION IN DESPATCHES (FOR GALLANT AND DISTINGUISHED SERVICE - POSTHUMOUS) | | | | | |
| BRONZE OAK LEAF EMBLEM | | | | | |
| CAPTAIN | | | | | |
| 39406 | 11.12.51 | 368568 | BUCHANAN | C.N.A. | Killed in action |
| MENTION IN DESPATCHES (FOR GALLANT AND DISTINGUISHED SERVICES) | | | | | |
| BRONZE OAK LEAF EMBLEM | | | | | |
| MAJOR | | | | | |
| 39484 | 04.03.52 | 58189 | GORDON-INGRAM | A.J. | |

| London Gazette issue No. | London Gazette date | Personal Service number | Name | Christian names | Remarks |
|---|---|---|---|---|---|
| MENTION IN DESPATCHES (FOR GALLANT AND DISTINGUISHED SERVICES) (continued) | | | | | |
| BRONZE OAK LEAF EMBLEM | | | | | |
| CAPTAINS | | | | | |
| 39328 | 07.09.51 | 360427 | COOKSON | A.J. | |
| 39328 | 07.09.51 | 370990 | CROWE | N.D.L. | |
| LIEUTENANT | | | | | |
| 39164 | 06.03.51 | 397871 | EDINGTON | J.R.R. | |
| SERGEANT | | | | | |
| 39164 | 06.03.51 | 14486821 | SMITH | R. | |
| LANCE-CORPORAL | | | | | |
| 39164 | 06.03.51 | 14463874 | MITCHELL | A. | |
| PRIVATES | | | | | |
| 39164 | 06.03.51 | 19032452 | FARQUHARSON | W. | |
| 39328 | 07.09.51 | 22308632 | ROBINSON | W.J. | |
| | | not listed | SATCHWELL | F. | |
| 39164 | 06.03.51 | 22167188 | SMITH | J. | |
| 39164 | 06.03.51 | 22186483 | WRIGHT | R. | King's Own Scottish Borderers attached |

## AWARDS CONFERRED BY THE PRESIDENT OF THE UNITED STATES OF AMERICA

| London Gazette issue No. | London Gazette date | Personal Service number | Name | Christian names | Remarks |
|---|---|---|---|---|---|
| SILVER STAR - POSTHUMOUS | | | | | |
| MAJOR | | | | | |
| | | 50980 | MUIR, VC | Kenneth | listed: killed in action |
| 2ND LIEUTENANT | | | | | |
| | | 407775 | BUCHANAN | M.D.W. | listed: killed in action |
| PRIVATE | | | | | |
| | | 22186756 | HILL | E. | listed: killed in action Hill 282 |
| SILVER STAR | | | | | |
| LIEUTENANT-COLONEL | | | | | |
| 39254 | 08.06.51 | 36735 | NEILSON, DSO | George Leslie | Officer Commanding |
| MAJOR | | | | | |
| 39254 | 08.06.51 | 249171 | GILLIES | James Blair | |
| CAPTAIN | | | | | |
| 39254 | 08.06.51 | 368568 | BUCHANAN | Charles Neill Anselan | listed: missing |

## SUPPORTING COMBATANT AND NON-COMBATANT CORPS UNITS

| London Gazette issue No. | London Gazette date | Personal Service number | Name | Christian names | Remarks |
|---|---|---|---|---|---|

## AWARDS CONFERRED BY THE PRESIDENT OF THE UNITED STATES OF AMERICA

### BRONZE STAR - POSTHUMOUS

SERGEANT

| London Gazette issue No. | London Gazette date | Personal Service number | Name | Christian names | Remarks |
|---|---|---|---|---|---|
| | | 14460096 | PIGG | E. | Extracted from regimental listing; shown as killed in action Hill 282 |

### BRONZE STAR

MAJORS

| London Gazette issue No. | London Gazette date | Personal Service number | Name | Christian names | Remarks |
|---|---|---|---|---|---|
| 39254 | 08.06.51 | 77925 | REITH | Ian Douglas McNeil | |
| 39254 | 08.06.51 | 73991 | STEWART, MBE, MC, TD | James Douglas | |

SERGEANTS

| London Gazette issue No. | London Gazette date | Personal Service number | Name | Christian names | Remarks |
|---|---|---|---|---|---|
| 39150 | 16.02.51 | 5437226 | O'SULLIVAN, MM | Joseph | |
| 39254 | 08.06.51 | 2979673 | ROBERTSON | James Crane | |

CORPORAL

| London Gazette issue No. | London Gazette date | Personal Service number | Name | Christian names | Remarks |
|---|---|---|---|---|---|
| 39254 | 08.06.51 | 3318780 | WALKER | John | |

LANCE-CORPORALS

| London Gazette issue No. | London Gazette date | Personal Service number | Name | Christian names | Remarks |
|---|---|---|---|---|---|
| 39150 | 16.02.51 | 14471420 | FAIRHURST | Joseph Levie | |
| 39150 | 16.02.51 | 14477378 | WARD | Harry Gordon | |

PRIVATE

| London Gazette issue No. | London Gazette date | Personal Service number | Name | Christian names | Remarks |
|---|---|---|---|---|---|
| 39150 | 16.02.51 | 21181514 | WATTS | William Temple | |

| London Gazette issue No. | London Gazette date | Personal Service number | Name | Christian names | Remarks |
|---|---|---|---|---|---|

ROYAL ARMY CHAPLAINS' DEPARTMENT

MBE: MEMBER OF THE ORDER OF THE BRITISH EMPIRE
(IN RECOGNITION OF GALLANT AND DISTINGUISHED SERVICES WHILST A PRISONER OF WAR IN NORTH KOREA)

| | | | | | |
|---|---|---|---|---|---|
| 40146 | 09.04.54 | 353777 | REVEREND DAVIES | Stanley James | |

MBE: MEMBER OF THE ORDER OF THE BRITISH EMPIRE

| | | | | | |
|---|---|---|---|---|---|
| 40036 | 04.12.53 | 160767 | REVEREND NICHOL, MC | Thomas James Trail | |
| 40036 | 04.12.53 | 414458 | REVEREND PETRY | James Patrick | |
| 39831 | 24.04.53 | 415034 | REVEREND PRESTON | Frederick Arnold | |
| 39831 | 24.04.53 | 88099 | REVEREND RHYS | Wallace Wyne Price | |

MENTION IN DESPATCHES (FOR GALLANT AND DISTINGUISHED SERVICES)
BRONZE OAK LEAF EMBLEM

| | | | | | |
|---|---|---|---|---|---|
| 39328 | 07.09.51 | 91617 | REVEREND RYAN | J. | |
| 39666 | 10.10.52 | 163326 | REVERENCE WILD | E.A. | |

AWARDS CONFERRED BY THE PRESIDENT OF THE UNITED STATES OF AMERICA

BRONZE STAR

| | | | | | |
|---|---|---|---|---|---|
| 40249 | 06.08.54 | 231867 | REVEREND McKINNON, MC | Campbell | |
| 39999 | 27.10.53 | 88099 | REVEREND RHYS, MBE | Wallace Wyne Price | |

| London Gazette issue No. | London Gazette date | Personal Service number | Name | Christian names | Remarks |
|---|---|---|---|---|---|
| ROYAL ARMY SERVICE CORPS | | | | | The Old Wagon Train; Ally Sloper's Cavalry Raised 1888 |
| OBE: OFFICER OF THE ORDER OF THE BRITISH EMPIRE | | | | | |
| LIEUTENANT-COLONELS | | | | | |
| 39528 | 29.04.52 | 157326 | CROSBY, MC | Michael George Marsh | CRASC 1st Comwel Division |
| 39831 | 24.04.53 | 378704 | GRANT, MBE | John MacKenzie | CRASC 1st Comwel Division |
| 40036 | 04.12.53 | 56667 | NOYES, MBE | Richard Ralph George Gower Godson | CRASC 1st Comwel Division |
| MAJORS | | | | | |
| 39328 | 07.09.51 | 345648 | EASTWOOD, MBE | Boris James | ← Unit not listed |
| 39328 | 07.09.51 | 375972 | HUNTER | Frank Alan Ritchie | ← |
| MBE: MEMBER OF THE ORDER OF THE BRITISH EMPIRE | | | | | |
| MAJORS | | | | | |
| 39666 | 10.10.52 | 191374 | McCLAREN | George Edward | ← |
| 40036 | 04.12.53 | 113325 | NEWBERRY-COBBETT | Guy Lovel | ← |
| 39666 | 10.10.52 | 354678 | PEARCE | Geoffrey Thomas | ← |
| CAPTAINS | | | | | |
| 39870 | 01.06.53 | 384996 | BONFIELD | Robert Gerald | ← |
| 39831 | 24.04.53 | 284663 | DOHERTY | Charles Francis | ← |
| 39666 | 10.10.52 | 248027 | JENKINS | Robert William | ← |
| WARRANT OFFICER CLASS I | | | | | |
| 39528 | 29.04.52 | S/54807 | BRYDGES | William | ← |
| WARRANT OFFICER CLASS II | | | | | |
| 39870 | 01.06.53 | S/5182955 | TOOGOOD | Ernest James | ← |
| MC: MILITARY CROSS | | | | | |
| 2ND LIEUTENANT | | | | | |
| 39442 | 18.01.52 | 414338 | PATTERSON | Hugh Norman Campbell | attached 1st Bn The King's Own Scottish Borderers |
| BEM: BRITISH EMPIRE MEDAL | | | | | |
| WARRANT OFFICER CLASS I | | | | | |
| 39831 | 24.04.53 | S/179716 | JOLLY | Leonard Alfred | ← |
| WARRANT OFFICER CLASS II | | | | | |
| 40036 | 02.12.53 | T/112705 | HOUSTON | Robert Tullis | ← |
| CORPORALS | | | | | |
| 39528 | 29.04.52 | T/6028520 | ANDREWS | John Richard | |
| 39870 | 01.06.53 | S/22524180 | FRASER | George Manson | ← |
| | | | | | ← Unit not listed |

| London Gazette issue No. | London Gazette date | Personal Service number | Name | Christian names | Remarks |
|---|---|---|---|---|---|
| MENTION IN DESPATCHES (FOR GALLANT AND DISTINGUISHED SERVICES) | | | | | |
| BRONZE OAK LEAF EMBLEM | | | | | |
| MAJORS | | | | | |
| 40036 | 04.12.53 | 108219 | BELL | D. | ⇷ Unit not listed |
| 39164 | 06.03.51 | 375972 | HUNTER | F.A.R. | ⇷ |
| 39328 | 07.09.51 | 143646 | MONCUR | G.W. | ⇷ |
| 39831 | 24.04.53 | 98866 | PARRY, MBE | R.K.M. | ⇷ |
| 39328 | 07.09.51 | 57075 | POTTER | F.H. | ⇷ |
| 39531 | 02.05.52 | 366192 | SHOTTER | G.B. | ⇷ |
| CAPTAINS | | | | | |
| 39831 | 24.04.53 | 187279 | AGGLETON | J.W. | ⇷ |
| 39831 | 24.04.53 | 155107 | BOON | J.B. | ⇷ |
| 39831 | 24.04.53 | 378921 | DESBOROUGH | T.H. | ⇷ |
| 39831 | 24.04.53 | 189316 | FINLAY | D. | ⇷ |
| 39328 | 07.09.51 | 259139 | SEATON | G.D. | ⇷ |
| 40036 | 04.12.53 | 376564 | TOWNSEND | N.F. | ⇷ |
| 39528 | 29.04.52 | 365525 | TURNER | D.G. | ⇷ |
| 39831 | 24.04.53 | 370080 | WEEKS | L.P. | ⇷ |
| 39528 | 29.04.52 | 170374 | WETHERALL | C.J. | ⇷ |
| LIEUTENANT | | | | | |
| 39870 | 01.06.53 | 393710 | JENKINS | R.E.L. | ⇷ |
| WARRANT OFFICERS CLASS I | | | | | |
| 40036 | 04.12.53 | S/57155 | DARIEN | T. | ⇷ |
| 39870 | 01.06.53 | T/110528 | VARLEY | L. | ⇷ |
| WARRANT OFFICERS CLASS II | | | | | |
| 40036 | 04.12.53 | T/45313 | FEE | W. | ⇷ |
| 39831 | 24.04.53 | S/10711786 | FELLOWS | J.G.L. | ⇷ |
| 39870 | 01.06.53 | S/5568518 | HINCH | H.F. | ⇷ |
| 39831 | 24.04.53 | T/14428504 | ROBINSON | C.R. | ⇷ |
| STAFF SERGEANTS | | | | | |
| 39870 | 01.06.53 | 22530019 | ASHFIELD | G.F. | ⇷ |
| 39831 | 24.04.53 | S/14459211 | IMMS | C.J. | ⇷ |
| 40036 | 04.12.53 | S/14417275 | MacCORMICK | N. | ⇷ |
| 39831 | 24.04.53 | 22773531 | PECKHAM | F.W. | ⇷ |
| 39831 | 24.04.53 | S/14927403 | WALLIS | R. | ⇷ |

⇷ Unit not listed

| London Gazette issue No. | London Gazette date | Personal Service number | Name | Christian names | Remarks |
|---|---|---|---|---|---|
| **MENTION IN DESPATCHES (FOR GALLANT AND DISTINGUISHED SERVICES)** | | | | | |
| **BRONZE OAK LEAF EMBLEM** | | | | | |
| **SERGEANTS** | | | | | |
| 40036 | 04.12.53 | T/22234176 | BERRY | J. | ← Unit not listed |
| 39528 | 29.04.52 | T/21016634 | BOURNE | W. | ← |
| 40036 | 04.12.53 | S/7012885 | CARTY | M. | ← |
| 39531 | 02.05.52 | T/2877111 | DONALD | J.M. | ← |
| 39528 | 29.04.52 | T/21183156 | HOWARD | E.J. | ← |
| 39831 | 02.05.53 | T/6984623 | MILLIGAN | K. | ← |
| 39531 | 02.05.52 | S/22285022 | PEPPERELL | L.K. | ← |
| 39870 | 01.06.53 | S/22231903 | SPEIRS | D. | ← |
| 40036 | 04.12.53 | T/22533130 | TURNBULL | J. | ← |
| 39831 | 24.04.53 | T/83110 | WAGGOTT | L. | ← |
| 39831 | 24.04.53 | S/5253029 | WILLETTS | F.N. | ← |
| **CORPORALS** | | | | | |
| 39666 | 10.10.52 | S/22425725 | BOYNE | I.F. | ← |
| 39398 | 30.11.51 | T/60092191 | JONES | F. | ← |
| 40036 | 04.12.53 | S/22588332 | METCALFE | B.H. | ← |
| 39831 | 24.04.53 | S/14458819 | PERRY | W.G. | ← |
| 39870 | 01.06.53 | S/22286402 | WARRILOW | S. | ← |
| **LANCE-CORPORAL** | | | | | |
| 40036 | 04.12.53 | T/22774300 | DUNNE | R. | ← |
| **PRIVATE** | | | | | |
| 39870 | 01.06.53 | 22588733 | EVISON | B.M. | ← |

## AWARDS CONFERRED BY THE PRESIDENT OF THE UNITED STATES OF AMERICA

| London Gazette issue No. | London Gazette date | Personal Service number | Name | Christian names | Remarks |
|---|---|---|---|---|---|
| **BRONZE STAR** | | | | | |
| **MAJORS** | | | | | |
| 39462 | 08.02.52 | 345648 | EASTWOOD, OBE | Boris James | ← |
| 39999 | 29.10.53 | 191734 | McLAREN, MBE | George Edward | ← |
| 39462 | 08.02.52 | 143646 | MONCUR | George Webster | ← |

← Unit not listed

| London Gazette issue No. | London Gazette date | Personal Service number | Name | Christian names | Remarks |
|---|---|---|---|---|---|
| ROYAL ARMY MEDICAL CORPS | | | | | Linseed Lancers; Poultice Wallopers<br>Raised 1898 |
| OBE: OFFICER OF THE ORDER OF THE BRITISH EMPIRE | | | | | |
| LIEUTENANT-COLONEL | | | | | |
| 39831 | 24.04.53 | 89376 | MARKS, MB | Richard Leonard | Officer Commanding 26th Field Ambulance |
| MBE: MEMBER OF THE ORDER OF THE BRITISH EMPIRE | | | | | |
| MAJORS | | | | | |
| 39528 | 29.04.52 | 378103 | MILN | William Gordon | ← Unit not listed |
| 39870 | 01.06.53 | 399883 | TUCKER, MRCS, LRCP | Desmond Keith | ← |
| MC: MILITARY CROSS | | | | | |
| MAJOR | | | | | |
| 39205 | 17.04.51 | 380058 | BOWEN, MRCS, LRCP | Cecil William | attached 1st Bn The Royal Northumberland Fusiliers |
| CAPTAINS | | | | | |
| 40036 | 04.12.53 | 359331 | HICKEY, MB, BS | Robert Patrick | attached 1st Bn The Gloucestershire Regiment |
| 40036 | 04.12.53 | 375425 | PATCHETT | Douglas Robert | attached 5th Royal Inniskilling Dragoon Guards |
| GM: GEORGE MEDAL | | | | | |
| CORPORAL | | | | | |
| 39756 | 20.01.53 | 22246864 | LOWES | Colin Leyshon | attached 61st Light Regiment Royal Artillery |
| MM: MILITARY MEDAL | | | | | |
| SERGEANT | | | | | |
| 39282 | 10.07.51 | 7265498 | BAKER | Bernard | attached 1st Bn The Royal Northumberland Fusiliers |
| CORPORAL | | | | | |
| 40036 | 08.12.53 | 7265709 | PAPWORTH | Cyril James | attached 1st Bn The Gloucestershire Regiment |
| MENTION IN DESPATCHES (FOR GALLANT AND DISTINGUISHED SERVICES WHILST A PRISONER OF WAR) | | | | | |
| BRONZE OAK LEAF EMBLEM | | | | | |
| CAPTAIN | | | | | |
| 40146 | 09.04.54 | 378046 | FERRIE, MB | A.M. | |
| MENTION IN DESPATCHES (FOR GALLANT AND DISTINGUISHED SERVICES) | | | | | |
| BRONZE OAK LEAF EMBLEM | | | | | |
| LIEUTENANT-COLONEL | | | | | |
| 39528 | 29.04.52 | 63253 | MacLENNAN, OBE, MB | A. | Officer Commanding 26th Field Ambulance |
| MAJOR | | | | | |
| 39273 | 29.06.51 | 128376 | MONTGOMERY, MC, MB | D.H.R. | ←<br>← Unit not listed |

| London Gazette issue No. | London Gazette date | Personal Service number | Name | Christian names | Remarks |
|---|---|---|---|---|---|
| MENTION IN DESPATCHES (FOR GALLANT AND DISTINGUISHED SERVICES) (continued) | | | | | |
| BRONZE OAK LEAF EMBLEM | | | | | |
| CAPTAINS | | | | | |
| 39528 | 29.04.52 | 375403 | BRETLAND, MB | P.M. | € Unit not listed |
| 39870 | 01.06.53 | 417766 | DUNCAN, MB | P.G.J. | € |
| 40036 | 08.12.53 | 422438 | MACKIE, MB | E.D. | € |
| 39831 | 24.04.53 | 291757 | RUDD, MB | D.H.C. | € |
| 39870 | 01.06.53 | 291757 | RUDD, MB | D.H.C. | € |
| 40036 | 08.12.53 | 421354 | TUTTY, MB | C.L. | € |
| LIEUTENANT QUARTERMASTER | | | | | |
| 40036 | 08.12.53 | 359814 | GRANT | W.H. | € |
| LIEUTENANT | | | | | |
| 39870 | 01.06.53 | 421547 | SCOTT, MB | M.J. | € |
| WARRANT OFFICERS CLASS II | | | | | |
| 40036 | 08.12.53 | 7265105 | DAVIS | H. | € |
| 39666 | 10.10.52 | 7263593 | McCARTHY | D. | € |
| 39870 | 01.06.53 | 7261352 | PIERREPONT | W. | € |
| SERGEANTS | | | | | |
| 40036 | 08.12.53 | 7264762 | ROBERTS | J.A. | € |
| 39831 | 24.04.53 | 14444245 | SMITH | N.O. | attached 1st Bn The Welch Regiment |
| CORPORALS | | | | | |
| 39749 | 09.01.53 | 22258357 | ALLEN | K.S. | attached 1st Bn The Black Watch |
| 39528 | 29.04.52 | 22306091 | BARRON | J.A. | attached 1st Bn The Leicestershire Regiment |
| 40036 | 04.12.53 | 7264697 | BRUTON | E.T. | € |
| 40206 | 18.06.54 | 7265680 | GEARY | F.E. | € |
| 39831 | 24.04.53 | 22524163 | HANCOCK | J. | attached 1st Bn The Royal Fusiliers |
| 39528 | 29.04.52 | 829012 | HOLMES | M.W. | attached 1st Bn The King's Shropshire Light Infantry |
| 40036 | 04.12.53 | 22562959 | MARSHALL | T.A. | € |
| 39589 | 04.07.52 | 22540839 | SHELLAM | R. | attached 5th Inniskilling Dragoon Guards |
| | | 22393786 | WILKINSON | R. | attached 1st Bn The Royal Northumberland Fusiliers |
| 39831 | 24.04.53 | 22424141 | WILLIAMS | J.R.P. | € |
| PRIVATES | | | | | |
| 39273 | 29.06.51 | 7264882 | BROOME | E. | € |
| 39282 | 10.07.51 | 7265082 | BROWN | W.R. | € |
| 39282 | 10.07.51 | 7265070 | ELLIOTT | M.T. | € |

€ Unit not listed

| London Gazette issue No. | London Gazette date | Personal Service number | Name | Christian names | Remarks |
|---|---|---|---|---|---|

## AWARDS CONFERRED BY THE PRESIDENT OF THE UNITED STATES OF AMERICA

SILVER STAR

MAJOR

| | | | | | |
|---|---|---|---|---|---|
| 39462 | 28.02.52 | 128376 | MONTGOMERY, MC | Dennis Hugh Robert | ← |

BRONZE STAR

LIEUTENANT-COLONEL

| | | | | | |
|---|---|---|---|---|---|
| 39462 | 08.02.52 | 63253 | MacLENNAN, OBE, MB | Alastair | Officer Commanding 26th Field Ambulance |

CAPTAIN

| | | | | | |
|---|---|---|---|---|---|
| 39999 | 27.10.53 | 415707 | JORY, MB | Harold Ian | ← |

← Unit not listed

| London Gazette issue No. | London Gazette date | Personal Service number | Name | Christian names | Remarks |
|---|---|---|---|---|---|
| ROYAL ARMY ORDNANCE CORPS | | | | | The Ordnance Raised 1881 |
| OBE: OFFICER OF THE ORDER OF THE BRITISH EMPIRE | | | | | |
| LIEUTENANT-COLONEL | | | | | |
| 39870 | 01.06.53 | 95693 | JACKSON | Brian Bannister | ← Unit not listed |
| MBE: MEMBER OF THE ORDER OF THE BRITISH EMPIRE | | | | | |
| MAJOR | | | | | |
| 40036 | 04.12.53 | 124793 | ATHILL | Geoffrey John | ← |
| CAPTAINS | | | | | |
| 39212 | 24.04.51 | 371859 | LOVE | Henry | ← |
| 39870 | 01.06.53 | 311483 | PAIN | Alexander William Edward | ← |
| WARRANT OFFICER CLASS I | | | | | |
| 39831 | 24.04.53 | 15001517 | CORBITT | Basil George Malcolm | ← |
| BEM: BRITISH EMPIRE MEDAL | | | | | |
| STAFF SERGEANT | | | | | |
| 39666 | 10.10.52 | 329792 | STAMPER | Eric | ← |
| PRIVATES | | | | | |
| 39212 | 24.04.51 | 5339455 | AVERY | Wilfred Hubert | ← |
| 39212 | 24.04.51 | 3393754 | MELLING | Harry | ← |
| 39212 | 24.04.51 | 22251643 | ROBERTSON | William | ← |
| MENTION IN DESPATCHES (FOR GALLANT AND DISTINGUISHED SERVICES) | | | | | |
| BRONZE OAK LEAF EMBLEM | | | | | |
| MAJORS | | | | | |
| 39870 | 01.06.53 | 386088 | ALLEN | N.F.T. | ← |
| 39528 | 29.04.52 | 235413 | ANDREWS | H.A. | ← |
| 39531 | 02.05.52 | 309563 | EMMETT, MBE | T.H. | ← |
| 39328 | 07.09.51 | 252133 | GUSCOTT, MBE | J. | ← |
| CAPTAINS | | | | | |
| 39666 | 10.10.52 | 380056 | BATESON | R. | ← |
| 39328 | 07.09.51 | 164626 | BROWN | J.W. | ← |
| 39528 | 29.04.52 | 265939 | LAMBERT | E.J. | ← |
| 40036 | 04.12.53 | 273429 | PEAT | C.W. | ← |
| 39831 | 24.04.53 | 311718 | STONE, MBE | A.W. | ← Unit not listed |
| 2ND LIEUTENANT | | | | | |
| 39164 | 06.03.51 | 397411 | WHITE | G.A. | Killed in action whilst attached to 1st Bn The Middlesex Regiment |

| London Gazette issue No. | London Gazette date | Personal Service number | Name | Christian names | Remarks |
|---|---|---|---|---|---|
| MENTION IN DESPATCHES (FOR GALLANT AND DISTINGUISHED SERVICES) (continued) | | | | | |
| BRONZE OAK LEAF EMBLEM | | | | | |
| WARRANT OFFICER CLASS I | | | | | |
| 39666 | 10.10.52 | 13068644 | STONE | F. | ← Unit not listed |
| STAFF QUARTERMASTER SERGEANT | | | | | |
| 40036 | 04.12.53 | 6985838 | DUGGAN | J.C.W. | ← |
| SERGEANTS | | | | | |
| 39831 | 24.04.53 | 22418297 | DERRING | E.T. | ← |
| 39666 | 10.10.52 | 22225887 | FEARNLEY | A. | ← |
| 40036 | 04.12.53 | 22563239 | WEAVER | P.R. | ← |
| CORPORAL | | | | | |
| 40036 | 04.12.53 | 22542127 | CHEYNE | G. | ← |

## AWARDS CONFERRED BY THE PRESIDENT OF THE UNITED STATES OF AMERICA

| | | | | | |
|---|---|---|---|---|---|
| SILVER STAR - POSTHUMOUS | | | | | |
| 2nd LIEUTENANT | | | | | |
| 39254 | 08.06.51 | 397411 | WHITE | Geoffrey Allan | Killed in action whilst attached to 1st Bn The Middlesex Regiment |
| BRONZE STAR | | | | | |
| MAJOR | | | | | |
| 39999 | 27.10.53 | 386088 | ALLEN | Norman Francis Theodore | ← Unit not listed |

| London Gazette issue No. | London Gazette date | Personal Service number | Name | Christian names | Remarks |
|---|---|---|---|---|---|
| CORPS OF ROYAL ELECTRICAL AND MECHANICAL ENGINEERS | | | | | Remy<br>Raised 1942 |
| OBE: OFFICER OF THE ORDER OF THE BRITISH EMPIRE | | | | | |
| LIEUTENANT-COLONEL | | | | | |
| 39666 | 10.10.52 | 124314 | GOOD, MBE | Herbert George | CREME 1st Comwel Division |
| MBE: MEMBER OF THE ORDER OF THE BRITISH EMPIRE | | | | | |
| MAJOR | | | | | |
| 40036 | 04.12.53 | 75113 | SYKES | Kenneth Musgrave | ← Unit not listed |
| CAPTAINS | | | | | |
| 39666 | 10.10.52 | 303461 | KELLY | John Dennis | ← |
| 39831 | 24.04.53 | 314969 | O'CALLAGHAN | Roy Percival | ← |
| BEM: BRITISH EMPIRE MEDAL | | | | | |
| WARRANT OFFICERS CLASS II | | | | | |
| 39870 | 01.06.53 | 7587329 | HILLS | Peter William | ← |
| 39666 | 10.10.52 | 7584511 | HOPKINS | James Reynolds | ← |
| STAFF SERGEANTS | | | | | |
| 39870 | 01.06.53 | 22514505 | DEGNAN | Laurence Kitchener | ← |
| 40036 | 04.12.53 | 819880 | ELLIS | Ernest Frederick Albert | ← |
| 40036 | 04.12.53 | 6343388 | RIXON | Frank | ← |
| 39870 | 01.06.53 | 215928 | WHEELER | Ralph Michael | ← |
| SERGEANT | | | | | |
| 39870 | 01.06.53 | 19054350 | GRAND | John Edward | ← |
| CORPORAL | | | | | |
| 40146 | 09.04.54 | 22305928 | MATHEWS | Robert Frank | ← listed: prisoner of war |
| MENTION IN DESPATCHES (FOR GALLANT AND DISTINGUISHED SERVICES) | | | | | |
| BRONZE OAK LEAF EMBLEM | | | | | |
| LIEUTENANT-COLONEL | | | | | |
| 40036 | 04.12.53 | 77378 | PALMER, MBE | P.G. | CREME 1st Comwel Division |
| MAJORS | | | | | |
| 40036 | 04.12.53 | 249002 | HODDER | R.A. | ← |
| 39328 | 07.09.51 | 125446 | MATTHEWS, MBE | J.R. | ← |
| CAPTAINS | | | | | |
| 40036 | 04.12.53 | 243541 | AITKEN | C.R.S.McL. | ← |
| 39666 | 10.10.52 | 334020 | BIRCH | R.R. | ← |
| 40036 | 04.12.53 | 367199 | FAGERLUND | K. | ← |
| | | | | | ← Unit not listed |

| London Gazette issue No. | London Gazette date | Personal Service number | Name | Christian names | Remarks |
|---|---|---|---|---|---|
| MENTION IN DESPATCHES (FOR GALLANT AND DISTINGUISHED SERVICES) (continued) | | | | | |
| BRONZE OAK LEAF EMBLEM | | | | | |
| CAPTAINS (continued) | | | | | |
| 39870 | 01.06.53 | 325049 | HAYMAN | B. | € Unit not listed |
| 39831 | 24.04.53 | 359375 | HYDE | J.R. | € |
| LIEUTENANTS | | | | | |
| 39528 | 29.04.52 | 387781 | CRIB, B.Sc. | A. | € |
| 39666 | 10.10.52 | 411873 | DIX, MBE | M.C. | € |
| WARRANT OFFICER CLASS I | | | | | |
| 39666 | 10.10.52 | 7591121 | COLLARD | A.W. | € |
| WARRANT OFFICERS CLASS II | | | | | |
| 39666 | 10.10.52 | 945269 | BROWN | L.F. | € |
| 40036 | 04.12.53 | 2322441 | EARL | H.P.E | € |
| 39831 | 24.04.53 | 2547586 | NORMAN | B.W. | € |
| STAFF SERGEANTS | | | | | |
| 39831 | 24.04.53 | 22514943 | HACKETT | E.A.C. | attached 1st Bn The Welch Regiment |
| 39531 | 02.05.52 | 7587679 | HILLIER | F.C. | € |
| 39831 | 24.04.53 | 13063578 | HOLLAND | G. | € |
| 39666 | 10.10.52 | 21001057 | MYERS | J. | € |
| 39831 | 24.04.53 | 2548036 | RAYNER | A. | € |
| SERGEANTS | | | | | |
| 40036 | 04.12.53 | 2548141 | GILES | K.B. | € |
| 39666 | 10.10.52 | 21022095 | GRAY | G.E. | € |
| 39666 | 10.10.52 | 22028510 | REED | R.E. | € |
| 39666 | 10.10.52 | 21125715 | RIDDINGTON | E.R. | € |
| 39666 | 10.10.52 | 22270261 | SERAPH | W.H. | € |
| CORPORALS | | | | | |
| 39528 | 29.04.52 | 22305800 | McCOMB | T.Z.T. | € |
| 39589 | 04.07.52 | 2549251 | NEWPORT | J.A. | attached 5th Inniskilling Dragoon Guards |
| LANCE-CORPORAL | | | | | |
| 39870 | 01.06.53 | 22541798 | SHIPLEY | P.J. | € |
| CRAFTSMEN | | | | | |
| 39666 | 10.10.52 | 21187358 | VICKERS | G.W. | € |
| 39528 | 29.04.52 | 22272547 | WARNER | J. | € |

€ Unit not listed

| London Gazette issue No. | London Gazette date | Personal Service number | Name | Christian names | Remarks |
|---|---|---|---|---|---|
| CORPS OF ROYAL MILITARY POLICE | | | | | Red Caps<br>Raised 1877 |
| BEM: BRITISH EMPIRE MEDAL | | | | | |
| SERGEANT | | | | | |
| 40036 | 04.12.53 | 22197930 | WILLIAMS | Kenneth Leslie | ← Unit not listed |
| MENTION IN DESPATCHES (FOR GALLANT AND DISTINGUISHED SERVICES)<br>BRONZE OAK LEAF EMBLEM<br>WARRANT OFFICERS CLASS I | | | | | |
| 39831 | 24.04.53 | 868646 | SMEDLEY | J.A. | ← |
| 39528 | 29.04.52 | 4389066 | STEAD | G.S.P. | ← |
| WARRANT OFFICER CLASS II | | | | | |
| 39328 | 07.09.51 | 14381961 | BOTTING | J.W. | ← |
| STAFF SERGEANT (CQMS) | | | | | |
| 39870 | 01.06.53 | 5110016 | HAWKESWORTH | J. | ← |
| SERGEANTS | | | | | |
| 40036 | 04.12.53 | 2548278 | HOLLOWAY | C.J. | 1st Comwel Division Provost Company |
| 39666 | 10.10.52 | 7684030 | HOLT | F. | ← |
| 39666 | 10.10.52 | 1443786 | WADHAMS | W.T. | ← |
| CORPORAL | | | | | |
| 39328 | 07.09.51 | 22513104 | PELLATT | G.T. | ← |
| LANCE-CORPORALS | | | | | |
| 39528 | 29.04.52 | 19044776 | BOND | I. | ← |
| 40036 | 04.12.53 | 22547451 | CLEMENTS | M.W. | ← Unit not listed |

| London Gazette issue No. | London Gazette date | Personal Service number | Name | Christian names | Remarks |
|---|---|---|---|---|---|
| MILTARY PROVOST STAFF CORPS | | | | | |
| MENTION IN DESPATCHES (FOR GALLANT AND DISTINGUISHED SERVICE) BRONZE OAK LEAF EMBLEM | | | | | |
| STAFF SERGEANT | | | | | |
| 40036 | 04.12.53 | 926214 | THORNBER | L. | |

| London Gazette issue No. | London Gazette date | Personal Service number | Name | Christian names | Remarks |
|---|---|---|---|---|---|
| ROYAL ARMY PAY CORPS | | | | | Quill Drivers Raised 1878 |
| MBE: MEMBER OF THE ORDER OF THE BRITISH EMPIRE | | | | | |
| CAPTAIN | | | | | |
| 40036 | 04.12.53 | 251259 | HODGKINSON | Charles Howard | ← Unit not listed |
| MENTION IN DESPATCHES (FOR GALLANT AND DISTINGUISHED SERVICES) BRONZE OAK LEAF EMBLEM | | | | | |
| MAJOR | | | | | |
| 39528 | 29.04.52 | 159819 | GRANT | D.E. | ← Unit not listed |

| London Gazette issue No. | London Gazette date | Personal Service number | Name | Christian names | Remarks |
|---|---|---|---|---|---|
| ROYAL ARMY EDUCATIONAL CORPS | | | | | Raised 1920 |
| MBE: MEMBER OF THE ORDER OF THE BRITISH EMPIRE | | | | | |
| MAJOR | | | | | |
| 39870 | 01.06.53 | 293009 | POUNDS | John Sydney Noel | ← Unit not listed |
| CAPTAIN | | | | | |
| 40036 | 04.12.53 | 364253 | EDWARDS | Graham Howell | ← |
| BEM: BRITISH EMPIRE MEDAL | | | | | |
| SERGEANT | | | | | |
| 39870 | 01.06.53 | 22165063 | McKIERNAN | Thomas | ← |
| MENTION IN DESPATCHES (FOR DISTINGUISHED SERVICES) BRONZE OAK LEAF EMBLEM | | | | | |
| SERGEANTS | | | | | |
| 39831 | 24.04.53 | 22028446 | AYRES | B.McI.L. | ← |
| 39666 | 10.10.52 | 22232757 | KENNEDY | B.D. | ← |
| | | | | | ← Unit not listed |

| London Gazette issue No. | London Gazette date | Personal Service number | Name | Christian names | Remarks |
|---|---|---|---|---|---|
| INTELLIGENCE CORPS | | | | | Raised 1940 |
| MM: MILITARY MEDAL | | | | | |
| STAFF SERGEANT | | | | | |
| 40036 | 04.12.53 | 5348543 | JACKSON, DCM | Clifford | € Unit not listed |
| MENTION IN DESPATCHES (FOR GALLANT AND DISTINGUISHED SERVICES) BRONZE OAK LEAF EMBLEM | | | | | |
| CAPTAIN | | | | | |
| 39528 | 29.04.52 | 251555 | CLARK | J.L. | € |
| WARRANT OFFICERS CLASS II | | | | | |
| 39328 | 07.09.51 | 824341 | PRAILL | L.M.H. | € |
| 40206 | 18.06.54 | 61045 | REED, MM | K. | € |
| SERGEANT | | | | | |
| 39831 | 24.04.53 | 21001023 | GRAY | R. | € |

## AWARD CONFERRED BY THE PRESIDENT OF THE UNITED STATES OF AMERICA

| London Gazette issue No. | London Gazette date | Personal Service number | Name | Christian names | Remarks |
|---|---|---|---|---|---|
| BRONZE STAR | | | | | |
| SERGEANT | | | | | |
| 39999 | 27.10.53 | 2576437 | WELLS | John Nicholas | € |

€ Unit not listed

| London Gazette issue No. | London Gazette date | Personal Service number | Name | Christian names | Remarks |
|---|---|---|---|---|---|

ARMY PHYSICAL TRAINING CORPS

BEM: BRITISH EMPIRE MEDAL
(FOR GALLANT AND DISTINGUISHED SERVICE WHILST A PRISONER OF WAR IN NORTH KOREA)

WARRANT OFFICER CLASS II

| London Gazette issue No. | London Gazette date | Personal Service number | Name | Christian names | Remarks |
|---|---|---|---|---|---|
| 40146 | 09.04.54 | 1874123 | STRONG | Frederick George | attached 1st Bn The Gloucestershire Regiment (bodyguard to Lieutenant-Colonel Carne during the Battle of the River Imjin April 1951) |

| London Gazette issue No. | London Gazette date | Personal Service number | Name | Christian names | Remarks |
|---|---|---|---|---|---|
| ARMY CATERING CORPS | | | | | Raised 1941 |
| MBE: MEMBER OF THE ORDER OF THE BRITISH EMPIRE | | | | | |
| CAPTAIN | | | | | |
| 39870 | 01.06.53 | 406579 | MARLOW | Humphrey Robert William David | ← Unit not listed |
| BEM: BRITISH EMPIRE MEDAL | | | | | |
| WARRANT OFFICER CLASS II | | | | | |
| 40036 | 04.12.53 | 5436972 | BURT | ALFRED | ← |
| MENTION IN DESPATCHES (FOR GALLANT AND DISTINGUISHED SERVICES) | | | | | |
| SERGEANTS | | | | | |
| 39831 | 24.04.53 | 3909094 | HACKMAN | W.E.C. | attached 1st Bn The Welch Regiment |
| 39831 | 24.04.53 | 1512929 | WESTRING | P. | ← |
| | | | | | ← Unit not listed |

ROYAL AIR FORCE

| London Gazette issue No. | London Gazette date | Personal Service number | Name | Christian names | Remarks |
|---|---|---|---|---|---|
| ROYAL AIR FORCE | | | | | |
| OBE: OFFICER OF THE ORDER OF THE BRITISH EMPIRE | | | | | |
| WING COMMANDER | | | | | |
| 40107 | 23.02.54 | 39292 | MACKENZIE, DFC | Donald | ∈ Unit not listed |
| BAR TO DFC: DISTINGUISHED FLYING CROSS | | | | | |
| WING COMMANDER | | | | | |
| 39323 | 31.08.51 | 33544 | Le CHEMINANT, DFC | Peter de Lacey | ∈ |
| SQUADRON LEADER | | | | | |
| 39205 | 17.04.51 | 44131 | PROCTOR, DFC | John Ernest | ∈ |
| FLIGHT LIEUTENANT | | | | | |
| 39833 | 24.04.53 | 81884 | BERGMAN, DFC | Vaclav | ∈ |
| DFC: DISTINGUISHED FLYING CROSS | | | | | |
| SQUADRON LEADERS | | | | | |
| 40107 | 19.02.54 | 59846 | FRANCIS | Harold Thomas | ∈ |
| 39505 | 01.04.52 | 60129 | HELME, AFC | James Michael | ∈ |
| 40010 | 06.11.53 | 131083 | ORMSTON | John Thornton | ∈ |
| FLIGHT LIEUTENANTS | | | | | |
| 40107 | 19.02.54 | 150069 | ARNAUD | Jean Raymond | ∈ |
| 40169 | 07.05.54 | 579588 | BALL | Brian James | attached 77 Fighter/Interceptor Squadron RAAF |
| 40107 | 19.02.54 | 1664660 | BENNETT | Melville Glyn | ∈ |
| 39553 | 30.05.52 | 199075 | BLYTH, AFC | Colin Ian | attached 77 Fighter/Interceptor Squadron RAAF |
| 39661 | 03.10.52 | 58993 | BRAND | Robert Stanley | ∈ |
| 39505 | 01.04.52 | 50304 | BRIDGE | Timothy Allman | ∈ |
| 40107 | 19.02.54 | 311507 | CHESWORTH | George Arthur | ∈ |
| 40010 | 06.11.53 | 55548 | CLARKE | Dennis Harry | ∈ |
| 39661 | 03.10.52 | 59708 | DAY | Lewis James | ∈ |
| 39661 | 03.10.52 | 59230 | DOUCHE | John Roland | ∈ |
| 39323 | 31.08.51 | 132759 | HOUTHEUSEN | Herbert Joseph | ∈ |
| 39870 | 26.05.53 | 164659 | LAIDLAY | Andrew Malcolm | ∈ |
| 39906 | 07.07.53 | 56713 | MELLERS | John | ∈ |
| 39879 | 02.06.53 | 58144 | NICHOLLS | John Moreton | ∈ |
| 39553 | 30.05.52 | 59609 | SCANNELL, AFC | Maxwell | attached 77 Fighter/Interceptor Squadron RAAF |

∈ Unit not listed

| London Gazette issue No. | London Gazette date | Personal Service number | Name | Christian names | Remarks |
|---|---|---|---|---|---|
| DFC: DISTINGUISHED FLYING CROSS (continued) | | | | | |
| FLIGHT LIEUTENANTS (continued) | | | | | |
| 40010 | 10.11.53 | 163127 | WELLS | Kenneth John | ← Unit not listed |
| 39891 | 19.06.53 | 57562 | WHITWORTH-JONES | Michael Edward | attached 77 Fighter/Interceptor Squadron RAAF |
| FLYING OFFICERS | | | | | |
| 40078 | 19.01.54 | 607080 | ARNOTT | Donald Aston | attached 77 Fighter/Interceptor Squadron RAAF |
| 39833 | 24.04.53 | 1395880 | LEE | Ralph Matthew | ← |
| MASTER ENGINEER | | | | | |
| 39323 | 31.08.51 | 520776 | LEDINGHAM | Loggie George | ← |
| DFM: DISTINGUISHED FLYING MEDAL | | | | | |
| FLIGHT SERGEANTS | | | | | |
| 40107 | 19.02.54 | 1697306 | JAMESON | Kenneth | ← |
| 39833 | 24.04.53 | 1946243 | REILLY | James Michael | ← |
| 39505 | 01.04.52 | 1079277 | TAYLOR | John Holt | ← |
| SERGEANTS | | | | | |
| 39871 | 01.06.53 | 4036568 | GRACE | Bryan | ← |
| 39661 | 03.10.52 | 3046142 | KINCH | Donald Graham | ← |
| 40107 | 19.02.54 | 4037705 | KITCHING | James Everley | ← |
| 39323 | 31.08.51 | 1625446 | McCOURT | Gerard John Francis | ← |
| 40010 | 06.11.53 | 579380 | TAIT | Eric Charles Webber | ← |
| BEM: BRITISH EMPIRE MEDAL | | | | | |
| SENIOR TECHNICIAN | | | | | |
| 40010 | 06.11.53 | 578443 | PATTRICK | Robert Victor | 1903 AOP Flight |
| SERGEANT | | | | | |
| 39833 | 24.04.53 | 576935 | CARR | Robert William | ← |
| MENTION IN DESPATCHES (FOR GALLANT AND DISTINGUISHED SERVICES) | | | | | |
| BRONZE OAK LEAF EMBLEM | | | | | |
| WING COMMANDER | | | | | |
| 39661 | 03.10.52 | 39292 | MACKENZIE, DFC | D. | ← |
| SQUADRON LEADER | | | | | |
| 39833 | 24.04.53 | 70196 | EAMES | P.F. | ← |

← Unit not listed

| London Gazette issue No. | London Gazette date | Personal Service number | Name | Christian names | Remarks |
|---|---|---|---|---|---|
| MENTION IN DESPATCHES (FOR GALLANT AND DISTINGUISHED SERVICES) (continued) | | | | | |
| BRONZE OAK LEAF EMBLEM | | | | | |
| FLIGHT LIEUTENANTS | | | | | |
| 39323 | 31.08.51 | 58121 | ALDERSMITH | M.F. | € Unit not listed |
| 39323 | 31.08.51 | 3035025 | BOSTON | G.A. | € |
| 39906 | 07.07.53 | 3039492 | CHANDLER | E.S. | attached 77 Fighter/Interceptor Squadron RAAF |
| 39838 | 24.04.53 | 56018 | CRAIG, DFC | R. | € |
| 39323 | 31.08.51 | 138834 | DANDEKER, DFC | K. | € |
| 39205 | 17.04.51 | 59708 | DAY | L.J. | € |
| 39323 | 31.08.51 | 152692 | EVANS | K.J. | € |
| 39661 | 03.10.52 | 152692 | EVANS | K.J. | € |
| 39661 | 03.10.52 | 501300 | HILLS | D.G.M. | attached 77 Fighter/Interceptor Squadron RAAF |
| 39906 | 07.07.53 | 3039659 | HOLMES | W.G. | attached 77 Fighter/Interceptor Squadron RAAF |
| 39205 | 17.04.51 | 149840 | HORSNELL | B.C. | € |
| 39205 | 17.04.51 | 59079 | HUNTER | D.M. | € |
| 39205 | 17.04.51 | 59587 | HUTCHISON | A.J. | € |
| 39661 | 03.10.52 | 180490 | JOHNSON, DFC | K.W. | € |
| 39205 | 17.04.51 | 158126 | LAIRD, DFC | P.E. | € |
| 39205 | 17.04.51 | 59445 | MARTINDALE | R.A. | € |
| 39205 | 17.04.51 | 169694 | NICHOLSON | D.J. | € |
| 39504 | 01.04.52 | 150799 | POYSER | J. | € |
| 40078 | 19.01.54 | 607094 | PRICE | J.W. | attached 77 Fighter/Interceptor Squadron RAAF |
| 39504 | 01.04.52 | 52758 | ROBERTS | E.L. | € |
| 40107 | 19.02.54 | 3039256 | ROBINSON | D.B. | € |
| 39323 | 31.08.51 | 114580 | SPROATES | G. | € |
| 39323 | 31.08.51 | 54996 | STONE | R.S. | € |
| 39833 | 24.04.53 | 579749 | SWALWELL | L.C. | € |
| 39661 | 03.10.52 | 55826 | WEBB | E.J. | € |
| 39504 | 01.04.52 | 51913 | WELSH | G.P. | € |
| 40107 | 19.02.54 | 193416 | WILDY | E.P. | € |
| FLYING OFFICERS | | | | | |
| 39892 | 19.06.53 | 607136 | BURLEY | B.M. | attached 77 Fighter/Interceptor Squadron RAAF |
| 39661 | 03.10.52 | 3034100 | CROFT | G.G.C. | € |

€ Unit not listed

| London Gazette issue No. | London Gazette date | Personal Service number | Name | Christian names | Remarks |
|---|---|---|---|---|---|
| MENTION IN DESPATCHES (FOR GALLANT AND DISTINGUISHED SERVICES) (continued) | | | | | |
| BRONZE OAK LEAF EMBLEM | | | | | |
| FLYING OFFICERS (continued) | | | | | |
| 40365 | 31.12.54 | 2316889 | CRUICKSHANK | O.M. | attached 77 Fighter/Interceptor Squadron RAAF |
| 39504 | 01.04.52 | 3110204 | GALPIN | D.N. | ∈ Unit not listed |
| 40107 | 19.02.54 | 3504882 | GILBERT | G.H. | ∈ |
| 39906 | 07.07.53 | 501336 | HOOGLAND | A.J. | attached 77 Fighter/Interceptor Squadron RAAF |
| 39661 | 03.10.52 | 1395880 | LEE | R.M. | ∈ |
| 39833 | 24.04.53 | 2360134 | O'BRIEN | T.P. | ∈ |
| | | 607094 | PRICE | J.W. | ∈ |
| 39504 | 01.04.52 | 2221798 | REA | P.B. | ∈ |
| 39833 | 24.04.53 | 56184 | ROSE | R.P.H. | ∈ |
| 40169 | 07.05.54 | 3124168 | SMITH | D.S. | attached 77 Fighter/Interceptor Squadron RAAF |
| MASTER ENGINEERS | | | | | |
| 40107 | 19.02.54 | 548778 | DAVIDSON | J. | ∈ |
| 39205 | 17.04.51 | 566271 | KNOTT | R.A. | ∈ |
| MASTER SIGNALLER | | | | | |
| 39323 | 31.08.51 | 611112 | GIBSON, DFM | J.W. | ∈ |
| MASTER GUNNER | | | | | |
| 39833 | 24.04.53 | 520178 | SANDERSON | W. | ∈ |
| FLIGHT SERGEANTS | | | | | |
| 40010 | 06.11.53 | 329942 | BAIRD | D. | ∈ |
| 40010 | 06.11.53 | 4016026 | BOOTH | R.J. | ∈ |
| 40010 | 06.11.53 | 1853391 | BOWN | J.A. | ∈ |
| 39504 | 01.04.52 | 1802625 | BURLTON | K.W. | ∈ |
| 39833 | 24.04.53 | 1258599 | D'SANTOS | P.E. | ∈ |
| 29504 | 01.04.52 | 576103 | FRASER | W.L. | ∈ |
| 39833 | 24.04.53 | 2227595 | IBBOTSON | A. | ∈ |
| 40107 | 19.02.54 | 578612 | NICHOLAS | R.J.K. | ∈ |
| 39661 | 03.10.52 | 1623957 | MATTHEWS | E. | ∈ |
| 39661 | 03.10.52 | 946349 | ROBERTSON | N.C. | ∈ |
| 39833 | 24.04.53 | 3031405 | TEDMAN | W.I. | ∈ |
| 39504 | 01.04.52 | 1319839 | TOLMAN | J.H. | ∈ |
| 39661 | 03.10.52 | 963676 | WEBB | E. | ∈ |

∈ Unit not listed

| London Gazette issue No. | London Gazette date | Personal Service number | Name | Christian names | Remarks |
|---|---|---|---|---|---|
| MENTION IN DESPATCHES (FOR GALLANT AND DISTINGUISHED SERVICES) (continued) | | | | | |
| BRONZE OAK LEAF EMBLEM | | | | | |
| SERGEANTS | | | | | |
| 39504 | 01.04.52 | 1890616 | ANDREETTI | L.L.C. | € Unit not listed |
| 39323 | 31.08.51 | 2238618 | ASHBROOK | G.A.H. | € |
| 39504 | 01.04.52 | 3054126 | BRUCE | G.H. | € |
| 40107 | 19.02.54 | 1798857 | CAHILL | J. | € |
| 39661 | 03.10.52 | 2222105 | COTTERILL | H. | € |
| 39661 | 03.10.52 | 4034666 | CROOK | H.J.B. | € |
| 40107 | 19.02.54 | 4041257 | CULVERHOUSE | M.J.B. | € |
| 39833 | 24.04.53 | 4034639 | CUNNINGHAM | F.McK. | € |
| 39833 | 24.04.53 | 2236628 | DODGE | D.M. | € |
| 40010 | 06.11.53 | 2211523 | EVANS | A. | € |
| 39504 | 01.04.52 | 1589488 | FOZARD | E. | € |
| 40010 | 06.11.53 | 3119444 | HAYWARD | M.P. | € |
| 39323 | 31.08.51 | 579262 | HOLLIDAY | W.G. | € |
| 39504 | 01.04.52 | 3046142 | KINCH | D.G. | € |
| 39323 | 31.08.51 | 1593503 | LEE | G.R. | € |
| 39833 | 24.04.53 | 4027355 | LEVER | J.A. | € |
| 39323 | 31.08.51 | 1836321 | LLEWELLYN | R.H. | € |
| 39661 | 03.10.52 | 3022888 | MANSON | D.J. | € |
| 40010 | 06.11.53 | 4036276 | REGAN | J.J. | € |
| 40010 | 06.11.53 | 3507274 | SCARLE | D.G. | € |
| 40010 | 06.11.53 | 1592919 | TAIT | R.E. | € |
| 39205 | 17.04.51 | 1850342 | WEAVER | F.H.Q. | € |
| 39323 | 31.08.51 | 2225613 | WORTHINGTON | A.R. | € |
| CORPORALS | | | | | |
| 39661 | 03.10.52 | 3500408 | COPPIN | A.C.R. | € |
| 40107 | 19.02.54 | 4026406 | ELLIOTT | L. | € |
| 40010 | 06.11.53 | 4026805 | FISH | D. | € |
| 40107 | 19.02.54 | 4030493 | GARRATT | H.E. | € |
| 40010 | 06.11.53 | 584223 | GROGAN | F.G. | € |
| 40010 | 06.11.53 | 592259 | IVINS | R. | € |
| 40107 | 19.02.54 | 1921653 | LONGE | E.H. | € |
| 40107 | 19.02.54 | 2467902 | MOULTON | T.P. | € |
| 39504 | 01.04.52 | 578724 | PARR | P.O. | € |
| 40107 | 19.02.54 | 4018901 | PRUDDEN | L.F. | € |

€ Unit not listed

| London Gazette issue No. | London Gazette date | Personal Service number | Name | Christian names | Remarks |
|---|---|---|---|---|---|
| **MENTION IN DESPATCHES (FOR GALLANT AND DISTINGUISHED SERVICES) (continued)** | | | | | |
| **BRONZE OAK LEAF EMBLEM** | | | | | |
| **SENIOR AIRCRAFTMEN** | | | | | |
| 39661 | 03.10.52 | 1922097 | ASHTON | A. | € Unit not listed |
| 40107 | 19.02.54 | 4056544 | BEYNON | R.A. | € |
| 39833 | 24.04.53 | 4042563 | CORRINGE | W.A. | € |
| 40107 | 19.02.54 | 4067665 | HARTSHORN | C.G. | € |
| 40010 | 06.11.53 | 4087486 | PUGSLEY | R.T. | € |

## AWARDS CONFERRED BY THE PRESIDENT OF THE UNITED STATES OF AMERICA

| London Gazette issue No. | London Gazette date | Personal Service number | Name | Christian names | Remarks |
|---|---|---|---|---|---|
| **LEGION OF MERIT, DEGREE OF COMMANDER** | | | | | |
| **AIR VICE MARSHAL** | | | | | |
| 40250 | 06.08.54 | not listed | BARNETT, CBE, DFC | Denis Hensley Fulton | € |
| **DISTINGUISHED FLYING CROSS** | | | | | |
| **SQUADRON LEADERS** | | | | | |
| 39531 | 02.05.52 | 409731 | ADDERLEY, AFC | Michael Charles | € |
| 39999 | 27.10.53 | 122337 | BALDWIN, DSO,DFC,AFC | John Robert | € listed: missing |
| 39999 | 27.10.53 | 46874 | DANIEL, DSO,DFC | Stephen Walter | € |
| **FLIGHT LIEUTENANTS** | | | | | |
| 40250 | 06.08.54 | 85764 | DICKINSON | Richard John Frederick | € |
| 40250 | 06.08.54 | 57728 | DUNLOP | Dennis Aichison | € |
| 40250 | 06.08.54 | 607070 | GORDON-JOHNSON | Ian | € |
| 40250 | 06.08.54 | 3039279 | GRANVILLE-WHITE | John Herbert | € |
| 40250 | 06.08.54 | 607020 | LOVELL | John Henry James | € |
| 40250 | 06.08.54 | 58468 | MAITLAND | John Ramsey | € |
| 39999 | 27.10.53 | 58144 | NICHOLLS, DFC | John Moreton | € |
| 40250 | 06.08.54 | 582051 | WATSON | Royston | € |
| **FLYING OFFICER** | | | | | |
| 40250 | 06.08.54 | 4038794 | SMITH | Brian Newbatt | € |
| **BRONZE STAR** | | | | | |
| **WING COMMANDER** | | | | | |
| 39326 | 25.05.51 | 41545 | BOXER, DSO, DFC | Alan Hunter Cachemaille | € |

€ Unit not listed

| London Gazette issue No. | London Gazette date | Personal Service number | Name | Christian names | Remarks |
|---|---|---|---|---|---|
| | | | | | |

## AWARDS CONFERRED BY THE PRESIDENT OF THE UNITED STATES OF AMERICA

| London Gazette issue No. | London Gazette date | Personal Service number | Name | Christian names | Remarks |
|---|---|---|---|---|---|
| BRONZE STAR (continued) | | | | | |
| SQUADRON LEADERS | | | | | |
| 39531 | 02.05.52 | 409731 | ADDERLEY, AFC | Michael Charles | € Unit not listed |
| 39999 | 27.10.53 | 36211 | De La PERRELLE | Victor Bretton | € |
| 39326 | 25.05.51 | 43475 | SACH, DFC, AFC | Jack Frederick | € |
| FLIGHT LIEUTENANT | | | | | |
| 40250 | 06.08.54 | 119209 | LAUGHTON | Leonard Stanley | € |
| FIRST OAK LEAF CLUSTER TO AIR MEDAL | | | | | |
| FLYING OFFICER | | | | | |
| 40250 | 06.08.54 | 4038794 | SMITH | Brian Newbatt | € |
| AIR MEDAL | | | | | |
| WING COMMANDERS | | | | | |
| 39326 | 25.05.51 | 41545 | BOXER, DSO, DFC | Alan Hunter Cachemaille | € |
| 39326 | 25.05.51 | 83267 | JOHNSON, DSO, DFC | James Edgar | € |
| 40250 | 06.08.54 | 74337 | MERIFIELD, DSO, DFC | John Roy Hugh | € |
| 39326 | 25.05.51 | 33211 | WYKEHAM-BARNES, DSO, OBE, DFC, AFC | Peter Guy | € |
| SQUADRON LEADERS | | | | | |
| 39531 | 02.05.52 | 40973 | ADDERLEY, AFC | Michael Charles | € |
| 39326 | 25.05.51 | 45720 | BODIEN, DSO, DFC | Henry Erskine | € |
| 39999 | 27.10.53 | 133668 | HARBISON | William | € |
| 40250 | 06.08.54 | 152607 | HIGSON | Ernest Maxwell | € |
| 39999 | 27.10.53 | 59129 | LELONG, DFC | Roy Emile | € |
| 40250 | 10.08.54 | 144393 | RANDLE, MBE, AFC, DFM | William Samuel Oliver | € |
| 39999 | 27.10.53 | 59609 | SCANNELL, DFC, AFC | Maxwell | attached 77 Fighter/Interceptor Squadron RAAF |
| 39999 | 27.10.53 | 178074 | SPRAGG, DFC | Brian Joseph | € |
| FLIGHT LIEUTENANTS | | | | | |
| 40250 | 06.08.54 | 607006 | BAYNE | Thomas Nigel Malcolm | € |
| 39999 | 27.10.53 | 199075 | BLYTHE, DFC, AFC | Colin Ian | attached 77 Fighter/Interceptor Squadron RAAF |
| 42250 | 06.08.54 | 2370100 | CHICK | John Francis Henry | € |
| 39531 | 02.05.52 | 46874 | DANIEL, DSO, DFC | Stephen Walter | € |

€ Unit not listed

| London Gazette issue No. | London Gazette date | Personal Service number | Name | Christian names | Remarks |
|---|---|---|---|---|---|
| | | | | | |

AWARDS CONFERRED BY THE PRESIDENT OF THE UNITED STATES OF AMERICA

AIR MEDAL

FLIGHT LIEUTENANTS (continued)

| | | | | | |
|---|---|---|---|---|---|
| 40250 | 06.08.54 | 178290 | DOWNES | Colin Bernard Walker | ∢ Unit not listed |
| 40250 | 06.08.54 | 57728 | DUNLOP | Dennis Aichison | ∢ |
| 39999 | 27.10.53 | 151649 | EASLEY | Francis | attached 77 Fighter/Interceptor Squadron RAAF |
| 40250 | 06.08.54 | 58042 | FRENCH | Roy | ∢ |
| 40250 | 06.08.54 | 607070 | GORDON-JOHNSON | Ian | ∢ |
| 40250 | 06.08.54 | 3039279 | GRANVILLE-WHITE | John Herbert | ∢ |
| 39999 | 27.10.53 | 58310 | KNIGHT | Rex | ∢ |
| 42050 | 06.08.54 | 607020 | LOVELL | John Harry James | ∢ |
| 40250 | 06.08.54 | 58017 | McELHAW | Timothy John | ∢ |
| 40250 | 06.08.54 | 58468 | MAITLAND | John Ramsey | ∢ |
| 40250 | 06.08.54 | 607091 | MANSELL | James Alec | ∢ |
| 40250 | 06.08.54 | 528355 | MURPHY | John Noel | ∢ |
| 40250 | 06.08.54 | 58147 | RYAN | James Anthony | ∢ |
| 40250 | 06.08.54 | 3046032 | SAWYER | Peter Gifford | ∢ |
| 40250 | 06.08.54 | 582051 | WATSON | Royston | ∢ |

FLYING OFFICERS

| | | | | | |
|---|---|---|---|---|---|
| 40250 | 06.08.54 | 4012760 | DEVINE | Colin David | ∢ |
| 40250 | 06.08.54 | 4038794 | SMITH | Brian Newbatt | ∢ |
| 40250 | 06.08.54 | 4034678 | WILKINSON | Geoffrey Crichton | ∢ |

SERGEANT

| | | | | | |
|---|---|---|---|---|---|
| 39999 | 27.10.53 | 1804890 | LAMB | Leslie Robert | attached 77 Fighter/Interceptor Squadron RAAF listed: deceased |

∢ Unit not listed

AWARDS FOR NON-OPERATIONAL SERVICES IN JAPAN IN CONNECTION WITH UNITED NATIONS DURING KOREAN WAR

| London Gazette issue No. | London Gazette date | Personal Service number | Name | Christian names | Remarks |
|---|---|---|---|---|---|
| ROYAL NAVY | | | | | |
| CBE: COMMANDER OF THE ORDER OF THE BRITISH EMPIRE | | | | | |
| CAPTAIN | | | | | |
| 39560 | 30.05.52 | not listed | VILLIERS | Robert Alexander | ← Appointment not listed |
| CAPTAIN (E) | | | | | |
| 39871 | 26.05.53 | not listed | GATEY | Colin | ← |
| OBE: OFFICER OF THE ORDER OF THE BRITISH EMPIRE | | | | | |
| COMMANDERS | | | | | |
| 39248 | 07.06.51 | not listed | GRAY | John Michael Dudgeon | ← |
| 39738 | 30.12.52 | not listed | PETER | Reginald Pendennis | ← |
| COMMANDER (S) | | | | | |
| 40058 | 01.01.54 | not listed | FENLEY | James Robert | ← |
| LIEUTENANT-COMMANDER (E) | | | | | |
| 39560 | 05.06.52 | not listed | GREEN | James Albert | ← |
| LIEUTENANT-COMMANDER (S) | | | | | |
| 39426 | 28.12.51 | not listed | FINLAY, MBE | Robert James Bell | ← |
| MBE: MEMBER OF THE ORDER OF THE BRITISH EMPIRE | | | | | |
| SENIOR COMMISSIONED COMMUNICATIONS OFFICER | | | | | |
| 39871 | 26.05.53 | not listed | BROOKS | Thomas Reginald | ← |
| LIEUTENANT-COMMANDER | | | | | |
| 39831 | 26.05.53 | not listed | MILLER | Austin Lawrence | ← |
| LIEUTENANT (S) | | | | | |
| 39109 | 29.12.50 | not listed | McGOLDRICK | Joseph Augustine | ← |
| BEM: BRITISH EMPIRE MEDAL | | | | | |
| CHIEF YEOMAN OF SIGNALS | | | | | |
| 40058 | 29.12.53 | D/JX136681 | LEYTHORNE, DSM | Thomas James | ← Vessel/Unit not listed |
| CHIEF PETTY OFFICER WRITER | | | | | |
| 39871 | 01.06.53 | P/MX58767 | PALMER | Kenneth Leslie | ← |
| CHIEF AIRCRAFT ARTIFICER | | | | | |
| 39738 | 01.01.53 | L/FX76697 | HOWLETT | Walter Arthur | ← |
| CHIEF PETTY OFFICER STORES | | | | | |
| 39109 | 29.12.50 | P/MX59634 | GARROD | Harry | ← |
| LEADING WRITER | | | | | |
| 39426 | 28.12.51 | P/SMX771303 | SORBIE | William | ← |
| YEOMAN OF SIGNALS | | | | | |
| 39560 | 30.05.52 | D/JX144680 | HUTCHINGS | William Kenneth | ← |

← Appointment/Vessel/Unit not listed

| London Gazette issue No. | London Gazette date | Personal Service number | Name | Christian names | Remarks |
|---|---|---|---|---|---|
| 41 INDEPENDENT COMMANDO, ROYAL MARINES | | | | | |
| BEM: BRITISH EMPIRE MEDAL | | | | | |
| MARINE (D) | | | | | |
| 39248 | 07.06.51 | CH/X118708 | BEEVOR | Fred | |

| London Gazette issue No. | London Gazette date | Personal Service number | Name | Christian names | Remarks |
|---|---|---|---|---|---|
| ARMY | | | | | |
| CBE: COMMANDER OF THE ORDER OF THE BRITISH EMPIRE | | | | | |
| BRIGADIER | | | | | |
| 39738 | 30.12.52 | 41082 | BATTEN, DSO, OBE | Richard Hutchison | Late Infantry |
| COLONEL | | | | | |
| 39871 | 26.05.53 | 161029 | BLUNDELL, OBE | Richard Victor | Royal Army Ordnance Corps |
| OBE: OFFICER OF THE ORDER OF THE BRITISH EMPIRE | | | | | |
| LIEUTENANT-COLONELS | | | | | |
| 39871 | 26.05.53 | 202079 | LOCKSLEY | Archibald Geddes | The Durham Light Infantry |
| 39738 | 30.12.52 | 38726 | LONSDALE, DSO | Maurice Robert | The North Staffordshire regiment |
| 40058 | 29.12.53 | 65320 | WRIGHT, DSO | David | Royal Army Medical Corps |
| MAJORS | | | | | |
| 39426 | 28.10.51 | 282169 | HEAD | James Richard | Royal Army Ordnance Corps |
| 39109 | 29.12.50 | 194885 | KNIGHT | Paul James Banks | The South Staffordshire Regiment |
| MBE: MEMBER OF THE ORDER OF THE BRITISH EMPIRE | | | | | |
| MAJORS | | | | | |
| 39560 | 30.05.52 | 73584 | BRAMWELL, MB | Stephen Oliver | Royal Army Medical Corps |
| 40058 | 29.12.53 | 70757 | CATON | Raymond March | The Queen's Royal Regiment |
| 39871 | 26.05.53 | 265764 | HEYDEN | Arthur Thomas John | Corps of Royal Engineers |
| 39871 | 26.05.53 | 143588 | MARTIN | John Samuel | The Welch Regiment |
| 39560 | 30.05.52 | 185024 | SMITH | James Christie | Corps of Royal Electrical and Mechanical Engineers |
| 39426 | 28.12.51 | 323338 | STOKES | Peter John Scott | The King's Royal Rifle Corps |
| CAPTAIN | | | | | |
| 40058 | 29.12.53 | 349265 | LAVIOLETTE | Hector Leslie | Royal Pioneer Corps |
| WARRANT OFFICER CLASS I | | | | | |
| 39738 | 30.12.52 | 7592456 | BARTLETT | Leonard Cecil | Corps of Royal Electrical and Mechanical Engineers |
| 39426 | 28.10.51 | 2656222 | THOMAS | Alfred | Coldstream Guards |
| 39738 | 30.12.52 | 10522132 | WITHERS | Denis | Royal Army Ordnance Corps |
| WARRANT OFFICER CLASS II | | | | | |
| 39871 | 26.05.53 | 3648689 | LOCKWOOD | James | Army Physical Training Corps |

| London Gazette issue No. | London Gazette date | Personal Service number | Name | Christian names | Remarks |
|---|---|---|---|---|---|
| BEM: BRITISH EMPIRE MEDAL | | | | | |
| WARRANT OFFICERS CLASS II | | | | | |
| 39738 | 01.01.53 | S/22533321 | BALDWIN | Thomas Alexander | Royal Army Service Corps |
| | | 22305665 | GARLAND | John Peace | Royal Army Ordnance Corps |
| | | 7591029 | GAY | Anthony | Corps of Royal Electrical and Mechanical Engineers |
| 40058 | 29.12.53 | 6200792 | HOCKLEY | George Frederick | The Middlesex Regiment |
| 39426 | 28.12.51 | 4547748 | JOYCE, MM | Austin | Welsh Guards |
| 39426 | 28.12.51 | 557144 | WALKER | Frederick Charles | 10th Royal Hussars |
| STAFF-SERGEANTS | | | | | |
| 39426 | 28.12.51 | 1115536 | BRIGGS | Frederick | Royal Army Ordnance Corps |
| 39426 | 28.12.51 | S/134372 | GIBBS | Robert | Royal Army Service Corps |
| 39871 | 26.05.53 | 4910198 | HART | John George | Royal Army Medical Corps |
| | | 14405259 | LITTLETON | Ambrose | Corps of Royal Electrical and Mechanical Engineers |
| 39871 | 26.05.53 | 5496033 | LUCAS | Ronald William | Royal Pioneer Corps |
| 39871 | 26.05.53 | 7589668 | PATERSON | Michael Purcell | Corps of Royal Electrical and Mechanical Engineers |
| 40058 | 29.12.53 | S/22285022 | PEPPERELL | Leonard Kenneth | Royal Army Service Corps |
| 39871 | 26.05.53 | 6397261 | POOLE | Charles George Smith | The South Staffordshire Regiment |
| 39738 | 01.01.53 | S/22285719 | RAWLINGS | Hubert Cyril | Royal Army Service Corps |
| COLOUR-SERGEANT | | | | | |
| | | 14427163 | ANDERSON | Charles Smith | The Gordon Highlanders |
| SERGEANTS | | | | | |
| 39738 | 01.01.53 | 22088798 | COMPTON | David Donald Dennis | Royal Army Ordnance Corps |
| 39738 | 01.01.53 | 5830376 | GOYMER | Frederick Libni | Royal Corps of Military Police |
| 39871 | 01.06.53 | 14958164 | POTTER | John Augustine | Royal Army Ordnance Corps |
| 40058 | 29.12.53 | 14193914 | SMITH | Kenneth Bernard | Corps of Royal Engineers |
| 39109 | 29.12.50 | 21182092 | TAYLOR | Walter James Ernest | Royal Army Service Corps |
| CORPORALS | | | | | |
| 39871 | 01.06.53 | 22541702 | COOMBES | Leslie Patrick | Royal Army Ordnance Corps |
| 40058 | 29.12.53 | 22621738 | SCOTT | Frederick John | Royal Army Ordnance Corps |

| London Gazette issue No. | London Gazette date | Personal Service number | Name | Christian names | Remarks |
|---|---|---|---|---|---|
| QUEEN ALEXANDRA'S ROYAL ARMY NURSING CORPS | | | | | |
| OBE: OFFICER OF THE ORDER OF THE BRITISH EMPIRE | | | | | |
| LIEUTENANT-COLONEL | | | | | |
| 39560 | 30.05.52 | 206521 | WIDGER, RRC | Phyllis | |
| ROYAL RED CROSS FIRST CLASS | | | | | |
| MAJORS | | | | | |
| 39560 | 30.05.52 | 206868 | CARSON | Jane | |
| 40058 | 29.12.53 | 213557 | INNES, OBE | Violet Mary | |
| 39871 | 26.05.53 | 206449 | STEWART | Priscilla Catherine | |
| ROYAL RED CROSS SECOND CLASS | | | | | |
| MAJOR | | | | | |
| 39738 | 01.01.53 | 206194 | HYNES | Florence Victoria | |

| London Gazette issue No. | London Gazette date | Personal Service number | Name | Christian names | Remarks |
|---|---|---|---|---|---|
| ROYAL AIR FORCE | | | | | |
| OBE: OFFICER OF THE ORDER OF THE BRITISH EMPIRE | | | | | |
| WING COMMANDER | | | | | |
| 39426 | 28.10.51 | 37464 | BURNSIDE, DSO, DFC | Dudley Henderson | |
| MBE: MEMBER OF THE ORDER OF THE BRITISH EMPIRE | | | | | |
| FLIGHT LIEUTENANT | | | | | |
| 39871 | 01.06.53 | 190480 | BISHOP | Walter John | |
| BEM: BRITISH EMPIRE MEDAL | | | | | |
| SERGEANT | | | | | |
| 39426 | 28.12.51 | 522340 | HAMPSON | Leslie Maurice | |
| CORPORALS | | | | | |
| 39871 | 01.06.53 | 2494489 | SKIPPER | Derek Arthur William | |
| 39871 | 01.06.53 | 579756 | SOWERBUTTS | Victor Frank | |

# COMMONWEALTH FORCES

CANADA

| London Gazette issue No. | London Gazette date | Personal Service number | Name | Christian names | Ship's title |
|---|---|---|---|---|---|
| ROYAL CANADIAN NAVY | | | | | |
| DSO: DISTINGUISHED SERVICE ORDER | | | | | |
| CAPTAIN | | | | | |
| 39420 | 28.12.51 | not listed | BROCK, DSC, CD | Jeffry Vanstone | HMCS Cayuga |
| OBE: OFFICER OF THE ORDER OF THE BRITISH EMPIRE | | | | | |
| CAPTAINS | | | | | |
| 40110 | 23.02.54 | not listed | LANDYMORE, CD | William Moss | HMCS Iroquois |
| 40096 | 09.02.54 | not listed | REED, DSC, CD | John Curwen | HMCS Athabaskan |
| COMMANDER | | | | | |
| 39657 | 30.09.52 | not listed | PLOMER, DSC, CD | James | HMCS Cayuga |
| BAR TO DSC: DISTINGUISHED SERVICE CROSS | | | | | |
| COMMANDER | | | | | |
| 39420 | 28.12.51 | not listed | WELLAND, DSC, CD | Robert Phillip | HMCS Athabaskan |
| DSC:DISTINGUISHED SERVICE CROSS | | | | | |
| CAPTAIN | | | | | |
| 39657 | 30.09.52 | not listed | TAYLOR, CD | Paul Dalrymple | HMCS Sioux |
| COMMANDERS | | | | | |
| 39657 | 30.09.52 | not listed | KING, CD | Dudley Gawen | HMCS Athabaskan |
| 40096 | 09.02.54 | not listed | LANTIER, CD | Dunn | HMCS Haida |
| 39540 | 13.05.52 | not listed | MADGWICK, CD | Edward Thomas George | HMCS Huron |
| 39725 | 23.12.52 | not listed | STEELE | Richard Miles | HMCS Nootka |
| LIEUTENANT-COMMANDERS | | | | | |
| 39881 | 05.06.53 | not listed | BOVEY | John Henry Gordon | HMCS Crusader |
| 39725 | 23.12.52 | not listed | SAXON, CD | Donald Roy | HMCS Cayuga |
| LIEUTENANTS | | | | | |
| 39420 | 28.12.51 | not listed | COLLIER | Andrew Laurence | HMCS Cayuga |
| 39881 | 05.06.53 | not listed | TUTTE, CD | Douglas Frederick | HMCS Iroquois |
| DSM:DISTINGUISHED SERVICE MEDAL | | | | | |
| CHIEF PETTY OFFICER (2) | | | | | |
| 39725 | 23.12.52 | 6030-H | BONNER, BEM | Albert Leo | HMCS Nootka |
| PETTY OFFICER | | | | | |
| 39881 | 05.06.53 | not listed | JAMIESON | Gerald Edwin | HMCS Iroquois |

| London Gazette issue No. | London Gazette date | Personal Service number | Name | Christian names | Ship's title |
|---|---|---|---|---|---|
| BEM: BRITISH EMPIRE MEDAL | | | | | |
| CHIEF PETTY OFFICER (1) | | | | | |
| 39420 | 28.12.51 | 21480-E | PEARSON | Douglas James | HMCS Cayuga |
| CHIEF PETTY OFFICER (2) | | | | | |
| 40096 | 09.02.54 | 31542 | VANDER-HAEGEN, DSM | George Charles | HMCS Athabaskan |
| PETTY OFFICER (2) | | | | | |
| 39420 | 28.12.51 | 6625-E | SHIELDS | Thomas | HMCS Athabaskan |
| 39540 | 13.05.52 | 4194-E | RANDALL | Edward Hannford | HMCS Nootka |
| MENTION IN DESPATCHES (FOR GALLANT AND DISTINGUISHED SERVICES - POSTHUMOUS) | | | | | |
| BRONZE OAK LEAF EMBLEM | | | | | |
| LIEUTENANT-COMMANDER | | | | | |
| | | not listed | QUINN | John Louis | HMCS Iroquois |
| MENTION IN DESPATCHES (FOR GALLANT AND DISTINGUISHED SERVICES) | | | | | |
| BRONZE OAK LEAF EMBLEM | | | | | |
| CAPTAINS | | | | | |
| 39285 | 13.07.51 | not listed | BROCK, DSC | Jeffry Vanstone | HMCS Cayuga |
| | | not listed | LANDYMORE, OBE, CD | William Moss | HMCS Iroquois |
| | | not listed | TAYLOR, DSC, CD | Paul Dalrymple | HMCS Sioux |
| COMMANDERS | | | | | |
| | | not listed | FRASER-HARRIS,DSC,CD | Alexander Beaufort Fraser | HMCS Nootka |
| 39285 | 13.07.51 | not listed | WELLAND, DSC, CD | Robert Phillip | HMCS Athabaskan |
| LIEUTENANT-COMMANDER (C) | | | | | |
| | | not listed | SHORTEN, CD | Harry | HMCS Athabaskan |
| LIEUTENANT-COMMANDER | | | | | |
| | | not listed | SAUNDERS | Frank Phillippo Rich | HMCS Nootka |
| SURGEON LIEUTENANT | | | | | |
| | | not listed | WEST | Chris Alfred | HMCS Athabaskan |
| ORDNANCE LIEUTENANT | | | | | |
| | | not listed | GIROUX | Gerald Joseph | HMCS Athabaskan |
| LIEUTENANT (L) | | | | | |
| | | not listed | BANFIELD | Nelson Ralph | HMCS Sioux |
| LIEUTENANT | | | | | |
| | | not listed | McCULLOCH | Paul Lancelot Steele | HMCS Athabaskan |
| COMMISSIONED GUNNER (TAS) | | | | | |
| | | not listed | HURL | David William | HMCS Sioux |
| CHIEF ENGINE ROOM ARTIFICER | | | | | |
| 39285 | 13.07.51 | 21673-E | DEAR | Edward Victor | HMCS Athabaskan |

| London Gazette issue No. | London Gazette date | Personal Service number | Name | Christian Names | Ship's title |
|---|---|---|---|---|---|
| MENTION IN DESPATCHES (FOR GALLANT AND DISTINGUISHED SERVICES) (continued) | | | | | |
| BRONZE OAK LEAF EMBLEM | | | | | |
| CHIEF PETTY OFFICERS | | | | | |
| | | not listed | BROWN | Harry Edward | HMCS Cayuga |
| | | not listed | CLARK | Lennox | HMCS Athabaskan |
| | | not listed | DAVIES | Ralph Evans | HMCS Cayuga |
| | | not listed | EWALD, CD | Frederick | HMCS Crusader |
| | | not listed | GOLD | Alfred | HMCS Crusader |
| | | not listed | LEARY | Joseph Ernest | HMCS Nootka |
| | | not listed | MEADS | John Leonard | HMCS Athabaskan |
| | | not listed | MORGAN, CD | Harry Cecil | HMCS Athabaskan |
| | | not listed | MOYES | William David | HMCS Athabaskan |
| | | not listed | SHEA | John Thornton | HMCS Athabaskan |
| | | not listed | VANTHAAF | George Edward | HMCS Sioux |
| | | not listed | WILLIAMS | Richard | HMCS Crusader |
| | | not listed | WINTER | reginal | HMCS Huron |
| PETTY OFFICERS | | | | | |
| | | not listed | FORTIN | Joseph Emilien Benoit | HMCS Nootka |
| | | not listed | SHAW | Samuel Henry | HMCS Athabaskan |
| | | not listed | SMITH | Ralph | HMCS Haida |
| LEADING SEAMAN | | | | | |
| | | not listed | ROBERTS | William John | HMCS Cayuga |
| ABLE SEAMAN | | | | | |
| | | not listed | STEWART | James Gordon | HMCS Crusader |

AWARDS CONFERRED BY THE PRESIDENT OF THE UNITED STATES OF AMERICA

| London Gazette issue No. | London Gazette date | Personal Service number | Name | Christian Names | Ship's title |
|---|---|---|---|---|---|
| LEGION OF MERIT, DEGREE OF COMMANDER | | | | | |
| COMMANDER | | | | | |
| | | not listed | MADGWICK, DSC, CD | Edward Thomas George | HMCS Huron |
| LEGION OF MERIT, DEGREE OF OFFICER | | | | | |
| CAPTAIN | | | | | |
| | | not listed | BROCK, DSO, DSC, CD | Jeffry Vanstone | HMCS Cayuga |
| COMMANDERS | | | | | |
| | | not listed | PLOMER, OBE, DSC, CD | James | HMCS Cayuga |
| | | not listed | TAYLOR, DSC, CD | Paul Dalrymple | HMCS Sioux |
| | | not listed | WELLAND, DSC, CD | Robert Phillip | HMCS Athabaskan |

| London Gazette issue No. | London Gazette date | Personal Service number | Name | Christian names | Ship's title |
|---|---|---|---|---|---|
| LEGION OF MERIT, DEGREE OF LEGIONNAIRE | | | | | |
| COMMANDERS | | | | | |
| | | not listed | FRASER-HARRIS,DSC,CD | Alexander Beaufort Fraser | HMCS Nootka |
| | | not listed | KING, DSC, CD | Dudley Gawen | HMCS Athabaskan |
| BRONZE STAR | | | | | |
| COMMANDER | | | | | |
| | | not listed | BOVEY, DSC, CD | John Henry Gordon | HMCS Crusader |
| DISTINGUISHED FLYING CROSS | | | | | |
| LIEUTENANT | | | | | |
| | | not listed | MacBRIEN | Joseph J. | Task Force 77 |

ROYAL CANADIAN ARMY

| London Gazette issue No. | London Gazette date | Personal Service number | Name | Christian names | Remarks |
|---|---|---|---|---|---|

HEADQUARTERS/STAFF

(INCLUDES ALL THOSE WHOSE UNITS ARE NOT SHOWN IN LONDON GAZETTE OR OFFICIAL CANADIAN LISTINGS)

CB: COMPANION OF THE ORDER OF THE BATH

BRIGADIER

| | | | | | |
|---|---|---|---|---|---|
| 39569 | 10.06.52 | not listed | ROCKINGHAM, CBE, DSO, ED | John Meredith | Commander 25 Canadian Infantry Brigade |

CBE: COMMANDER OF THE ORDER OF THE BRITISH EMPIRE

BRIGADIERS

| | | | | | |
|---|---|---|---|---|---|
| 39819 | 07.04.53 | ZP1378 | BOGERT, DSO, OBE, CD | Mortimer Patrick | Commander 25 Canadian Infantry Brigade |
| | | not listed | FLEURY, MBE, ED | Frank James | Commander Military Mission Far East |

OBE: OFFICER OF THE ORDER OF THE BRITISH EMPIRE

LIEUTENANT-COLONEL

| | | | | | |
|---|---|---|---|---|---|
| | | not listed | SARE | Paul Francis Lionel | Assistant Adjutant General, Canadian Military Mission Far East |

MBE: MEMBER OF THE ORDER OF THE BRITISH EMPIRE

MAJORS

| | | | | | |
|---|---|---|---|---|---|
| 39819 | 07.04.53 | ZK984 | BAKER, CD | Alfred Jeffery | HQ 25 Infantry Brigade |
| 39885 | 12.06.53 | ZB1125 | CLANCY, MC | John Anthony | HQ 1st Commonwealth Division |
| 39622 | 12.08.52 | ZH419 | HAMILTON | C. John Alexander | HQ 25 Infantry Brigade |
| 39819 | 07.04.53 | ZP1995 | LEACH, CD | John Edward | HQ 25 Infantry Brigade |
| 39980 | 02.10.53 | ZP1584 | McDONALD, CD | Thomas Murray | HQ 25 Infantry Brigade (also shown as MacDONALD) |
| 40102 | 16.02.54 | SB2864 | MAYER, CD | Paul Augustus | HQ 1st Commonwealth Division |

BEM: BRITISH EMPIRE MEDAL

SERGEANT

| | | | | | |
|---|---|---|---|---|---|
| | | not listed | TUTTE | Kenneth Gordon | L of C & Base Troops - BCFK |

MENTION IN DESPATCHES (FOR GALLANT AND DISTINGUISHED SERVICES)

BRONZE OAK LEAF EMBLEM

CAPTAIN

| | | | | | |
|---|---|---|---|---|---|
| | | not listed | STAPLES | R.J. | Liaison Officer BCFK |

| Canadian/ London Gazette issue No. | Canadian/ London Gazette date | Personal Service number | Name | Christian names | Remarks |
|---|---|---|---|---|---|
| MENTION IN DESPATCHES (FOR GALLANT AND DISTINGUISHED SERVICES) (continued) | | | | | |
| BRONZE OAK LEAF EMBLEM | | | | | |
| LIEUTENANTS | | | | | |
| 40 | 03.10.53 | ZC3214 | BEEMAN | William John Milton | HQ 25 Infantry Brigade |
| 45 | 08.11.52 | ZK3641 | CORRY | Geoffrey Donald | HQ 25 Infantry Brigade |
| PRIVATES | | | | | |
| | | not listed | COLLIER | Vehsey Eric | ∊ Unit not listed |
| | | not listed | HENRIKSON | Harold Richard | ∊ Unit not listed |
| 31 | 02.08.52 | SB7819 | RUTTAN | Lawrence Albert | ∊ Unit not listed: infantry |

AWARDS CONFERRED BY THE PRESIDENT OF THE UNITED STATES OF AMERICA

| Canadian/ London Gazette issue No. | Canadian/ London Gazette date | Personal Service number | Name | Christian names | Remarks |
|---|---|---|---|---|---|
| LEGION OF MERIT, DEGREE OF OFFICER | | | | | |
| BRIGADIERS | | | | | |
| | | not listed | ALLARD, CBE, DSO, ED | Jean Victor | Commander 25 Canadian Infantry Brigade |
| | | not listed | BOGERT, CBE, DSO, CD | Mortimer Patrick | Commander 25 Canadian Infantry Brigade |
| | | not listed | FLEURY, CBE, ED | Frank James | Commander Military Mission Far East |
| | | not listed | ROCKINGHAM, CB, CBE, DSO | John Meredith | Commander 25 Canadian Infantry Brigade |
| LEGION OF MERIT, DEGREE OF LEGIONNAIRE | | | | | |
| LIEUTENANT-COLONEL | | | | | |
| | | not listed | DANBY, DSO, OBE | Ernest Deighton | ∊ Unit not listed |
| BRONZE STAR | | | | | |
| LIEUTENANT-COLONEL | | | | | |
| | | ZF650 | AMY, DSO, OBE, MC | Edward Alfred Charles | RCAC - HQ 1st Commonwealth Division |
| MAJOR | | | | | |
| | | not listed | MEDLAND, DSO | Richard Dillon | ∊ Unit not listed: RCIC |
| CAPTAIN | | | | | |
| | | not listed | STAPLES | R.J. | Liaison Officer BCFK |
| DISTINGUISHED FLYING CROSS | | | | | |
| LIEUTENANT | | | | | |
| | | not listed | WARD | William Ernest | ∊ Unit not listed |

| Canadian/ London Gazette issue No. | Canadian London Gazette date | Personal Service number | Name | Christian names | Remarks |
|---|---|---|---|---|---|
| ROYAL CANADIAN HORSE ARTILLERY | | | | | |
| DSO: DISTINGUISHED SERVICE ORDER | | | | | |
| LIEUTENANT-COLONEL | | | | | |
| 39819 | 07.04.53 | ZP1754 | McNAUGHTON, CD (LESLIE) | Edward Murray Dalziel | 1st Regiment (This officer changed his name from McNaughton to Leslie in March 1953 |
| CBE: OFFICER OF THE ORDER OF THE BRITISH EMPIRE | | | | | |
| LIEUTENANT-COLONELS | | | | | |
| 39518 | 18.0452 | ZK107 | BAILEY, DSO, MBE, ED | Anthony John Beswick | 2nd Regiment |
| 39693 | 11.11.52 | ZP1379 | BROOKS, DSO, CD | Edward Geoffrey | 2nd Regiment |
| MBE: MEMBER OF THE ORDER OF THE BRITISH EMPIRE | | | | | |
| MAJORS | | | | | |
| 39819 | 07.04.53 | ZB9677 | COPCUTT | David Russell | 1st Regiment |
| 39980 | 06.10.53 | ZP1266 | ROBINSON, MC | Aaron | 1st Regiment |
| WARRANT OFFICER CLASS I (RSM) | | | | | |
| 39693 | 11.11.52 | SP9466 | ARMISHAW | Robert Vincent | 2nd Regiment |
| 39980 | 06.10.53 | SP9892 | SEED | William Thomas | 1st Regiment |
| WARRANT OFFICER CLASS II (RQMS) | | | | | |
| 39885 | 09.06.53 | SP6650 | HARDON | William Henry | 1st Regiment |
| MC: MILITARY CROSS | | | | | |
| CAPTAINS | | | | | |
| 39693 | 11.11.52 | ZF5016 | BERTHIAUME | Eric William | 2nd Regiment |
| 39819 | 07.04.53 | ZP2192 | CALDWELL, CD | Douglas Samuel | 1st Regiment (FOO A Battery A Troop) |
| 39819 | 07.04.53 | ZC4936 | HOWITT | Gerald Henry | 1st Regiment |
| LIEUTENANT | | | | | |
| 39518 | 18.04.52 | ZL4076 | O'BRENNAN | Matthew Terrance | 2nd Regiment |
| MM: MILITARY MEDAL | | | | | |
| SERGEANTS | | | | | |
| 39819 | 07.04.53 | SH60643 | GRAVELINE | William George | 1st Regiment |
| 39885 | 12.06.53 | SL934 | PROCUIK | Michael | 1st Regiment |
| BOMBARDIER | | | | | |
| 39622 | 12.08.52 | D800406 | DEARDEN | Thomas Edward | 2nd Regiment |
| LANCE-BOMBARDIERS | | | | | |
| 39518 | 18.04.52 | SD800167 | DORMAN | Francis Merton | 2nd Regiment |
| 39354 | 19.10.51 | C800133 | KING | Alan Osbourne | 2nd Regiment |

| Canadian/ London Gazette issue No. | Candian/ London Gazette date | Personal Service number | Name | Christian names | Remarks |
|---|---|---|---|---|---|
| MM: MILITARY MEDAL (continued) | | | | | |
| GUNNERS | | | | | |
| 39354 | 09.10.51 | D801275 | GARAUGHTY | Arthur Martin | 2nd Regiment |
| 39354 | 09.10.51 | L800032 | WISHART | Kenneth Wilfred | 2nd Regiment |
| BEM:BRITISH EMPIRE MEDAL | | | | | |
| BOMBARDIER | | | | | |
| 39885 | 01.06.53 | SL4093 | LONG | Harvey Eugene | 1st Regiment |
| MENTION IN DESPATCHES (FOR GALLANT AND DISTINGUISHED SERVICES) | | | | | |
| BRONZE OAK LEAF EMBLEM | | | | | |
| MAJORS | | | | | |
| 45 | 08.11.52 | TF70506 | GILLIS | Duncan Hugh | 2nd Regiment |
| 40 | 03.10.53 | ZP1480 | KEELER | Gerald Edward | 1st Regiment |
| 40 | 03.10.53 | ZM1971 | MURDOCH, MC | Ernest St John Charles | 1st Regiment |
| 40 | 03.10.53 | ZP1813 | OWEN, CD | Donald Maxwell | 1st Regiment |
| 45 08 | 08.11.52 | ZB4152 | PINKERTON | Samuel Morrison | 2nd Regiment |
| CAPTAINS | | | | | |
| 40 | 06.10.51 | ZA465 | CROWE, MC | John Douglas | 2nd Regiment |
| 45 | 08.11.52 | ZE252 | HOWARD | John Henry | 2nd Regiment (also shown as RCA) |
| 40 | 06.10.51 | ZK9686 | McGILLIVRAY | Charles Allan | 2nd Regiment |
| 45 | 08.11.52 | ZB4078 | MOLDAVER | Jack Isaac | 2nd Regiment |
| 45 | 08.11.52 | ZC3411 | STODDART | William Bateman | 2nd Regiment |
| 15 | 12.04.52 | KZ5066 | TIERNEY | Angus McDougall | 2nd Regiment |
| LIEUTENANT | | | | | |
| | | not listed | SMYTH | Robert Dunlop | 1st Regiment |
| WARRANT OFFICER CLASS II | | | | | |
| 45 | 08.11.52 | SD13525 | MALCOLM | Robert | 2nd Regiment |
| SERGEANTS | | | | | |
| 45 | 08.11.52 | SC850308 | MacKINNON | Donald Hillman | 2nd Regiment |
| 45 | 08.11.52 | SD800057 | McCORMICK | George Edward | 2nd Regiment |
| 31 | 02.08.52 | H800111 | WINKWORTH | Alfred Victor | 2nd Regiment (also shown as GNR. RCA) |
| BOMBARDIER | | | | | |
| 13 | 28.03.53 | SM100423 | REID | Gavin Ian | 1st Regiment |
| LANCE-BONBARDIERS | | | | | |
| | | not listed | BARTER | Walter Franklin | 1st Regiment |
| 15 | 12.04.52 | L800038 | BIRD | Gerald Everett | 2nd Regiment |
| 45 | 08.11.52 | SB7236 | BOLTON | Russell Wilburn | 2nd Regiment (also shown as 1st Bn RCR) |

| Canadian/ London Gazette issue No. | Canadian/ London Gazette date | Personal Service number | Name | Christian names | Remarks |
|---|---|---|---|---|---|
| MENTION IN DESPATCHES (FOR GALLANT AND DISTINGUISHED SERVICES) (continued) | | | | | |
| BRONZE OAK LEAF EMBLEM | | | | | |
| GUNNERS | | | | | |
| | | not listed | MONK | Cecil Clifford | 1st Regiment |
| 7 | 01.06.53 | SN1276 | SIMMONS | Gerald Joseph | 1st Regiment |
| 13 | 28.03.53 | SF8992 | WHITE | Daniel | 1st Regiment (also listed as WIGHT) |

## AWARDS CONFERRED BY THE PRESIDENT OF THE UNITED STATES OF AMERICA

| | | | | | |
|---|---|---|---|---|---|
| BRONZE STAR | | | | | |
| BOMBARDIER | | | | | |
| | | not listed | REID | Gavin Ian | 1st Regiment |
| AIR MEDAL | | | | | |
| CAPTAIN | | | | | |
| | | ZE252 | HOWARD | John Henry | 2nd Regiment (also listed as RCA) |

| Canadian/ London Gazette issue No. | Canadian/ London Gazette date | Personal Service number | Name | Christian names | Remarks |
|---|---|---|---|---|---|
| ROYAL CANADIAN ARMOURED CORPS - LORD STRATHCONA'S HORSE | | | | | |
| DSO: DISTINGUISHED SERVICE ORDER | | | | | |
| MAJOR | | | | | |
| 39693 | 11.11.52 | ZP1659 | JEWKES, MC | Victor Wilfred | Officer Commanding C Squadron (also listed as Victor William) |
| OBE: OFFICER OF THE ORDER OF THE BRITISH EMPIRE | | | | | |
| LIEUTENANT-COLONEL | | | | | |
| 39980 | 06.10.53 | ZF650 | AMY, DSO, MC | Edward Alfred Charles | HQ 1st Commonwealth Division |
| MBE: MEMBER OF THE ORDER OF THE BRITISH EMPIRE | | | | | |
| MAJOR | | | | | |
| 39819 | 07.04.53 | ZG2951 | ROXBURGH | John Sutton | Officer Commanding B Squadron (also listed as ROXBOROUGH) |
| WARRANT OFFICER CLASS II (SSM) | | | | | |
| 39980 | 06.10.53 | SP3602 | ARMER | Eric John | B Squadron |
| MC: MILITARY CROSS | | | | | |
| MAJOR | | | | | |
| 40102 | 16.02.54 | ZC2774 | ELLIS, CD | William Hodgson | Officer Commanding A Squadron |
| MM: MILITARY MEDAL | | | | | |
| SERGEANT | | | | | |
| 39646 | 16.09.52 | SF800236 | ALLEN | Trevor | C Squadron |
| TROOPER | | | | | |
| 39726 | 23.12.52 | SH61937 | STEVENSON | Roy Charles | B Squadron |
| BEM: BRITISH EMPIRE MEDAL | | | | | |
| SQUADRON QUARTERMASTER SERGEANT | | | | | |
| 39518 | 18.04.52 | SH237 | EVELEIGH | Douglas Fred | C Squadron |
| TROOPER | | | | | |
| 39819 | 07.04.53 | SM9445 | WYATT | Henry | B Squadron |
| MENTION IN DESPATCHES (FOR GALLANT AND DISTINGUISHED SERVICES) | | | | | |
| BRONZE OAK LEAF EMBLEM | | | | | |
| LIEUTENANTS | | | | | |
| 15 | 12.04.52 | ZK4389 | GLENDINNING | David Lorne | C Squadron |
| | | not listed | MacDONALD | Strathcona Clifton | C Squadron |
| 13 | 28.03.53 | ZB3601 | RUTHERFORD | Bruce Coleman | ← Unit not listed |
| 45 | 08.11.52 | ZG9788 | SMITH | Douglas Thompson | ← Unit not listed |

| Canadian/ London Gazette issue No. | Canadian/ London Gazette date | Personal Service number | Name | Christian names | Remarks |
|---|---|---|---|---|---|
| **MENTION IN DESPATCHES (FOR GALLANT AND DISTINGUISHED SERVICES) (continued)** | | | | | |
| **BRONZE OAK LEAF EMBLEM** | | | | | |
| **SERGEANTS** | | | | | |
| 15 | 12.04.52 | SF45980 | FORBES, MM | James Floyd | C Squadron |
| | | not listed | HARNOIS | Gerard Joseph | B Squadron |
| 7 | 01.06.53 | SB153051 | JACK | Walter Daniel | ← Unit not listed |
| 40 | 03.10.53 | SD118385 | STEWART | Alexander | ← |
| | | not listed | THWAITES | George Herbert | ← |
| **CORPORAL** | | | | | |
| 15 | 12.04.52 | H800326 | HARRISON | Sydney | C Squadron |
| **LANCE-CORPORAL** | | | | | |
| 7 | 13.02.54 | SD4749 | WEIR | Duncan Roland St John | RCD HQ 1st Commonwealth Division |
| **TROOPERS** | | | | | |
| 40 | 03.10.53 | SK9723 | FAIRLEY | Alexander | ← |
| 7 | 13.02.54 | SM9920 | HOLLICK | Albert William | A Squadron |
| 13 | 28.03.53 | SH61920 | OLYNYK | William | ← |
| **QUEEN'S COMMENDATION FOR BRAVE CONDUCT** | | | | | |
| **BRONZE OAK LEAF EMBLEM** | | | | | |
| **TROOPER** | | | | | |
| | | not listed | EASTER | Donald Carroll | B Squadron |

## BELGIAN AWARD

| Canadian/ London Gazette issue No. | Canadian/ London Gazette date | Personal Service number | Name | Christian names | Remarks |
|---|---|---|---|---|---|
| **CHEVALIER DE L'ORDRE DE LA COURONNE AVEC PALME AND LA CROIX DE GUERRE 1940 AVEC PALME** | | | | | |
| **LIEUTENANT** | | | | | |
| | | not listed | BULL | Robert Wallace | ← |

← Unit not listed

| Canadian/ London Gazette issue No. | Canadian/ London Gazette date | Personal Service number | Name | Christian names | Remarks |
|---|---|---|---|---|---|
| ROYAL CANADIAN ARTILLERY | | | | | |
| DSO: DISTINGUISHED SERVICE ORDER | | | | | |
| LIEUTENANT-COLONEL | | | | | |
| 40102 | 16.02.54 | ZP1977 | STERNE, MBE, CD | Henry William | Commander 81st Field Regiment |
| MBE: MEMBER OF THE ORDER OF THE BRITISH EMPIRE | | | | | |
| MAJOR | | | | | |
| 40102 | 16.02.54 | ZF1950 | BEER | John Pope | 81st Field Regiment |
| CAPTAIN | | | | | |
| 39980 | 06.10.53 | ZA1873 | HAUSER, CD | Rollin John | 81st Field Regiment |
| MC:MILITARY CROSS | | | | | |
| CAPTAIN | | | | | |
| 39759 | 23.01.53 | ZM2699 | de HART | John Edward | 81st Field Regiment |
| LIEUTENANT | | | | | |
| 39981 | 06.10.53 | ZF3843 | RUFFEE | George Edward Moodie | 81st Field Regiment |
| DFC: DISTINGUISHED FLYING CROSS | | | | | |
| CAPTAIN | | | | | |
| 40012 | 10.11.53 | ZK3344 | TEES | Peter Joseph Angwyn | 1903 Indep. Air Op Flight RAF |
| MENTION IN DESPATCHES (FOR GALLANT AND DISTINGUISHED SERVICES) | | | | | |
| BRONZE OAK LEAF EMBLEM | | | | | |
| MAJOR | | | | | |
| | | not listed | WIRMIER | Albert Lloyd | 81st Field Regiment (also listed as WEIRMIER) |
| BOMBARDIER | | | | | |
| 40 | 03.10.53 | SK6025 | REID | William | 81st Field Regiment (also listed as REED) |
| LANCE-BOMBARDIER | | | | | |
| | | not listed | WALSH | Robert Allen | 81st Field Regiment |
| GUNNERS | | | | | |
| | | not listed | SNOW | George Henry | ← Unit not listed |
| 31 | 02.08.52 | H800111 | WINKWORTH | Alfred Victor | ← Unit not listed (also shown as Sgt with 2nd Regiment RCHA) |

| Canadian/ London Gazette issue No. | Canadian/ London Gazette date | Personal Service number | Name | Christian names | Remarks |
|---|---|---|---|---|---|

AWARD CONFERRED BY THE PRESIDENT OF THE UNITED STATES OF AMERICA

AIR MEDAL

CAPTAIN

| | | | | | |
|---|---|---|---|---|---|
| | | not listed | HOWARD | John Henry | ⋠ Unit not listed (also shown as 2nd Regiment RCHA) |

BELGIAN AWARD

OFFICIER DE L'ORDRE DE LÉOPOLD II PALME AND LA CROIX DE GUERRE 1940 AVEC PALME

MAJORS

| | | | | | |
|---|---|---|---|---|---|
| | | not listed | STEWART | James Crossley | 81st Field Regiment |
| | | not listed | THERIAULT, MC | Joseph Elphége Yvan | 81st Field Regiment |

| Canadian/ London Gazette issue No. | Canadian/ London Gazette date | Personal Service number | Name | Christian names | Remarks |
|---|---|---|---|---|---|
| ROYAL CANADIAN CORPS OF ENGINEERS | | | | | |
| OBE: OFFICER OF THE ORDER OF THE BRITISH EMPIRE | | | | | |
| MAJOR | | | | | |
| 39518 | 10.04.52 | ZB450 | ROCHESTER | Donald Harvey | 57 Independent Field Squadron (also listed as Ronald Harvey) |
| MBE: MEMBER OF THE ORDER OF THE BRITISH EMPIRE | | | | | |
| MAJORS | | | | | |
| 39693 | 11.11.52 | ZF640 | ABBOTT | Albert Joseph | HQ 1st Commonwealth Division |
| 39693 | 11.11.52 | ZK603 | BALL | Harold William | 57 Independent Field Squadron and 28 Field Engineer Regiment |
| 39980 | 06.10.53 | ZG2116 | BLACK | Reay Melbourne | HQ 1st Commonwealth Division |
| 39885 | 12.06.53 | ZB495 | GALWAY, MC, GM | Edward Thomas | 23 Field Squadron |
| MC: MILITARY CROSS | | | | | |
| MAJOR | | | | | |
| 40102 | 16.02.54 | ZP1868 | SCHMIDLIN, MBE, CD | Laurence Edmond Carson | 59 Field Squadron |
| LIEUTENANTS | | | | | |
| 39981 | 06.10.53 | ZC4687 | CARTER | Charles David | 59 Field Squadron |
| 39518 | 18.04.52 | ZA800 | FREEBORN | Frederick Roberts | 57 Independent Field Squadron |
| MM: MILITARY MEDAL | | | | | |
| LANCE-CORPORAL | | | | | |
| | | SB13332 | RAPLEY | Gordon Stanley | 59 Field Squadron (also listed: RAPELEY) |
| BEM: BRITISH EMPIRE MEDAL | | | | | |
| SERGEANTS | | | | | |
| 39819 | 07.04.53 | SB144511 | GARDINER | Lorne Percival | 23 Field Squadron |
| 39819 | 07.04.53 | SM8932 | TOMELIN | Paul James | 25 Canadian Public Relations Unit |
| LANCE- CORPORAL | | | | | |
| 39491 | 14.03.52 | M800366 | McCREARY | Herbert John | 57 Independent Field Squadron |
| MENTION IN DESPATCHES (FOR GALLANT AND DISTINGUISHED SERVICES) | | | | | |
| BRONZE OAK LEAF EMBLEM | | | | | |
| CAPTAINS | | | | | |
| 7 | 13.02.54 | ZK3514 | McPHERSON | Angus Alexander | 59 Field Squadron (also listed as 59 Field Dental Unit) |
| 45 | 08.11.52 | ZB9619 | MITCHELL | Frederick Oliver | 57 Independent Field Squadron |

| Canadian/ London Gazette issue No. | Canadian/ London Gazette date | Personal Service number | Name | Christian names | Remarks |
|---|---|---|---|---|---|
| MENTION IN DESPATCHES (FOR GALLANT AND DISTINGUISHED SERVICES) (continued) | | | | | |
| BRONZE OAK LEAF EMBLEM | | | | | |
| LIEUTENANTS | | | | | |
| 45 | 08.11.52 | TK99548 | COOKE | Norman Edward | 57 Independent Field Squadron |
| 15 | 12.04.52 | TC38514 | KEATING | John Hugh | 25 & 57 Independent Field Squadrons |
| 40 | 03.10.53 | ZL3268 | PENNY | Richard Stanley | 57 Independent Field Squadron (also listed as 59 Indep. Field Squadron) |
| SERGEANT | | | | | |
| 45 | 08.11.52 | B801929 | STEELE | Wilbert Ross | 57 Independent Field Squadron |
| CORPORAL | | | | | |
| 13 | 28.03.53 | SB153495 | REID | Roy Hamilton | 25 Independent Field Squadron (also listed as 23 Indep. Field Squadron) |
| LANCE-CORPORALS | | | | | |
| 40 | 03.10.53 | SB13332 | RAPLEY | Gordon Stanley | 59 Field Squadron |
| 13 | 28.03.53 | SB34148 | WEBBER | Robert William Charles | 23 Field Squadron |
| SAPPERS | | | | | |
| 15 | 12.04.52 | SK14352 | FRIESEN | Henry Kenneth | 57 Independent Field Squadron |
| 7 | 01.06.53 | SM471 | HAWICK | William Dent | 23 Field Squadron |
| 13 | 28.03.53 | SD4870 | NORTON | John Henry | 23 Field Squadron |

| Canadian/ London Gazette issue No. | Canadian/ London Gazette date | Personal Service number | Name | Christian names | Remarks |
|---|---|---|---|---|---|
| ROYAL CANADIAN CORPS OF SIGNALS | | | | | |
| MBE: MEMBER OF THE ORDER OF THE BRITISH EMPIRE | | | | | |
| MAJOR | | | | | |
| 39622 | 12.08.52 | ZL116 | GEORGE, MC | Donald Harry | HQ 25 Infantry Brigade Signals |
| CAPTAINS | | | | | |
| 39885 | 09.06.53 | ZU2884 | CONNELL, CD | John Rossiter | HQ 1st Commonwealth Division Signals |
| 39693 | 11.11.52 | ZP2757 | WHEELER, CD | Walter Edward | HQ 25 Infantry Brigade Signals |
| MC: MILITARY CROSS | | | | | |
| LIEUTENANT | | | | | |
| 39908 | 10.07.53 | ZH3541 | CÔTÉ | Laurie George | att 3rd Bn Royal Canadian Regiment |
| BEM: BRITISH EMPIRE MEDAL | | | | | |
| SERGEANT | | | | | |
| 39980 | 06.10.53 | SB27275 | THOMPSON | Albert Edward | att 1st Regiment RCHA |
| MENTION IN DESPATCHES (FOR GALLANT AND DISTINGUISHED SERVICES) | | | | | |
| BRONZE OAK LEAF EMBLEM | | | | | |
| CAPTAINS | | | | | |
| 7 | 01.06.53 | ZB882 | DOIDGE | Ernest Karl | att 1st Regiment RCHA |
| 40 | 30.10.53 | ZB9841 | ELLIOTT | David Arthur | att 3rd Bn PPCLI |
| 45 | 08.11.52 | ZH2918 | HARRIS | Frederick Thomas | att 1st Bn PPCLI |
| 45 | 08.11.52 | ZP1794 | La ROSE, CD | Fernand Gaston | att 2nd Bn R22e Regiment |
| 7 | 13.02.54 | ZP2756 | SPICOLUK, CD | Frank John | att 3rd Bn RCR |
| LIEUTENANTS | | | | | |
| 45 | 08.11.52 | TD3878 | SORENSON | Frank Edward | att 2nd Regiment RCHA |
| 80 | 30.10.53 | ZM4329 | VANN | Gordon | att 81st Field Regiment RCA |
| WARRANT OFFICER CLASS II | | | | | |
| 15 | 12.04.52 | SD116748 | WINTERSON | Stanley Alexander | HQ 25 Infantry Brigade Signals |
| SERGEANT | | | | | |
| 45 | 08.11.52 | SA2402 | GUY | Harold John | att 2nd Regiment RCHA |
| CORPORALS | | | | | |
| 13 | 28.03.53 | SB154062 | BUCKNER | Edward Allen | HQ 25 Infantry Brigade Signals |
| 40 | 06.10.51 | SC850431 | CARRIERE | Albert Robert Joseph | ← Unit not listed |
| 40 | 06.10.51 | SC850052 | DISLEY | Joseph Anthony | ← Unit not listed |
| 40 | 03.10.53 | SF38669 | FLEMMING | Arthur George | HQ 25 Infantry Brigade Signals |
| 7 | 01.06.53 | SH20893 | McELREA | John Borland | att 1st Regiment RCHA |
| 7 | 01.06.53 | SB7878 | SEARLS | Harvey William | HQ 25 Infantry Brigade Signals |

| Canadian/ London Gazette issue No. | Canadian/ London Gazette date | Personal Service number | Name | Christian names | Remarks |
|---|---|---|---|---|---|
| MENTION IN DESPATCHES (FOR GALLANT AND DISTINGUISHED SERVICES) (continued) | | | | | |
| BRONZE OAK LEAF EMBLEM | | | | | |
| LANCE-CORPORAL | | | | | |
| 45 | 08.11.52 | D801830 | RODNEY | Peter Clovis | att 2nd Regiment RCHA |
| SIGNALMAN | | | | | |
| 45 | 08.11.52 | not listed | MANN | Arnold Rudolph | att 2nd Regiment RCHA |

| London Gazette issue No. | London Gazette date | Personal Service date | Name | Christian names | Remarks |
|---|---|---|---|---|---|
| ROYAL CANADIAN REGIMENT | | | | | |
| DSO: DISTINGUISHED SERVICE ORDER | | | | | |
| LIEUTENANT-COLONEL | | | | | |
| 39819 | 07.04.53 | Z1537 | BINGHAM | Peter Richard | Officer Commanding 1st Bn |
| OBE: OFFICER OF THE ORDER OF THE BRITISH EMPIRE | | | | | |
| LIEUTENANT-COLONELS | | | | | |
| 40102 | 16.02.54 | ZB406 | CAMBELL, MBE, CD | Kenneth Laidlaw | Officer Commanding 3rd Bn |
| 39693 | 11.11.52 | ZK4142 | CORBOULD, DSO, ED | Gordon Charleston | Officer Commanding 2nd Bn |
| 39518 | 18.04.52 | ZH106 | KEANE, DSO | Robert Angus | Officer Commanding 2nd Bn |
| MBE: MEMBER OF THE ORDER OF THE BRITISH EMPIRE | | | | | |
| MAJORS | | | | | |
| | | ZB1125 | CLANCY, MC | John Anthony | 1st Bn |
| 40102 | 16.02.54 | ZP2328 | COUCHE, CD | Richard Arthur | 3rd Bn |
| 39819 | 07.04.53 | ZP1751 | KLENAVIC | Francis | 1st Bn |
| MC: MILITARY CROSS | | | | | |
| MAJOR | | | | | |
| 39819 | 07.04.53 | ZF857 | TAYLOR, ED | George Gray | 1st Bn |
| CAPTAINS | | | | | |
| 39759 | 23.01.53 | ZP2810 | CLOUTIER, CD | Herbert George | 1st Bn |
| 40102 | 16.02.54 | ZM644 | JENKINS, CD | John Gallington | 3rd Bn |
| LIEUTENANTS | | | | | |
| 39759 | 23.01.53 | ZD4331 | CLARK | John | 1st Bn |
| 39709 | 02.12.52 | ZC4309 | GARDNER | Herbert Russell | 1st Bn |
| 39819 | 07.04.53 | ZC4693 | KING | Andrew Martin | 1st Bn |
| 39885 | 12.06.53 | ZK4730 | LOOMIS | Daniel Gordon | 1st Bn |
| 39467 | 12.02.52 | ZB4391 | MASTRONARDI | Edward John | 2nd Bn |
| 39627 | 19.08.52 | ZD3947 | PETERSON | Allan Angus Sloss | 1st Bn |
| 2ND LIEUTENANT | | | | | |
| 39908 | 10.07.53 | ZB10022 | HOLLYER | Edgar Herbert | 3rd Bn |
| DCM: DISTINGUISHED CONDUCT MEDAL | | | | | |
| PRIVATE | | | | | |
| 39518 | 18.04.52 | F800160 | BAUER | Rupert Edward | 2nd Bn |

| Canadian/ London Gazette issue No. | Canadian/ London Gazette date | Personal Service number | Name | Christian names | Remarks |
|---|---|---|---|---|---|
| MM: MILITARY MEDAL | | | | | |
| WARRANT OFFICERS CLASS II | | | | | |
| 39885 | 12.06.53 | SL1589 | FOX | George Maurice | 1st Bn |
| 39819 | 07.04.53 | SB42527 | JOHNSON | Leo Austin | 1st Bn |
| STAFF SERGEANT | | | | | |
| 39819 | 07.04.53 | SB41694 | ANDERSON | Peter | 1st Bn |
| SERGEANT | | | | | |
| 39759 | 23.01.53 | SC135658 | ENRIGHT | Gerald Emerson Peter | 1st Bn |
| CORPORALS | | | | | |
| 39709 | 02.12.52 | SC6603 | FOWLER | Karl Edmund | 1st Bn |
| 39610 | 29.07.52 | SN1106 | Le MOINE | Donald George | 1st Bn |
| 39908 | 10.07.53 | SF97294 | McNEIL | Joseph Cecil | 3rd Bn |
| 39601 | 18.07.52 | SD800398 | McORMOND | Kenneth Victor | 2nd Bn |
| 39819 | 07.04.53 | SN1051 | PELLEY | Cecil Wilbert Hoskin | 1st Bn |
| 39908 | 10.07.53 | SC800055 | PERO | William Daniel | 3rd Bn |
| 39622 | 12.08.52 | G800403 | SCOTT | Arthur Allan | 2nd Bn |
| 39725 | 19.08.52 | SG9631 | STINSON | Arthur Irvine | 1st Bn |
| PRIVATES | | | | | |
| 39408 | 14.12.51 | A800042 | BELL | Curtis Ora | 2nd Bn |
| 39622 | 12.08.52 | C800037 | CARLEY | Douglas Wesley | 2nd Bn |
| 39622 | 12.08.52 | D800577 | JOHNSON | Jack David | 2nd Bn |
| 39908 | 10.07.53 | SF6207 | JULIEN | George Patrick | 3rd Bn |
| 39433 | 04.01.52 | B800203 | PUGH | Wilfred Denis | 2nd Bn |
| 39408 | 14.12.51 | B810683 | ROWDEN | Gordon George | 2nd Bn |
| 39354 | 09.10.51 | C800000 | SARGEANT | John Archibald | 2nd Bn |
| BEM: BRITISH EMPIRE MEDAL | | | | | |
| SERGEANTS | | | | | |
| 39980 | 06.10.53 | SA63307 | ROSS | Ronald Leon | 1st Bn |
| 39980 | 06.10.53 | SA109040 | WALTERS | Wilbert Ernest | 1st Bn |
| LANCE-CORPORAL | | | | | |
| 39819 | 07.04.53 | SB7363 | NIXON | Melville Joseph | 1st Bn |
| MENTION IN DESPATCHES (FOR GALLANT AND DISTINGUISHED SERVICES) | | | | | |
| BRONZE OAK LEAF EMBLEM | | | | | |
| MAJORS | | | | | |
| 13 | 28.03.53 | ZC741 | COHEN, MBE | Elliot Lapetus | 1st Bn |
| 7 | 13.02.54 | ZG332 | FAIRWEATHER, MC | Arthur Thomas Edwin | 3rd Bn |
| 13 | 28.03.53 | ZA221 | HOLMES | Donald Ernest | 1st Bn |

| Canadian/ London Gazette issue No. | Canadian/ London Gazette date | Personal Service number | Name | Christian names | Remarks |
|---|---|---|---|---|---|
| MENTION IN DESPATCHES (FOR GALLANT AND DISTINGUISHED SERVICES) (continued) | | | | | |
| BRONZE OAK LEAF EMBLEM | | | | | |
| MAJORS (continued) | | | | | |
| 45 | 08.11.52 | TA37619 | PATERSON, DSO | John Frederick | 2nd Bn (also listed as PETERSON, DSO) |
| 13 | 28.11.53 | ZP1314 | RICHARDS, CD | Robert Stanley | 1st Bn |
| CAPTAINS | | | | | |
| 40 | 03.10.53 | ZP2608 | HENRY | Ronald Ormston | 1st Bn |
| 13 | 28.11.53 | ZF657 | MAHAR | Robert Herbert | 1st Bn |
| | | not listed | MULLIN | Merier Joseph | 3rd Bn |
| LIEUTENANTS | | | | | |
| 40 | 03.10.53 | ZA10016 | CASSAN | Edward Maxwell | 1st Bn |
| 45 | 08.11.52 | ZA4318 | DARLING | Carman Grant | 2nd Bn |
| 7 | 01.06.53 | ZD3809 | FAIRHEAD | Harold George | 1st Bn |
| 45 | 08.11.52 | TA32161 | HAMEL | John Starr | 2nd Bn |
| 45 | 08.11.52 | ZD4097 | HARTNETT | James Joseph | 2nd Bn (also listed as 2nd Bn PPCLI) |
| 45 | 08.11.52 | XF3763 | MacLEAN | James William | 1st Bn |
| 45 | 08.11.52 | TB14292 | McLELLAN | Alister Porteous | 2nd Bn (also listed as McLEMMAN) |
| 7 | 13.02.54 | ZD10138 | REEVES | Kenneth Joseph William | 3rd Bn |
| WARRANT OFFICER CLASS I (RSM) | | | | | |
| 40 | 03.10.53 | SP15427 | BURNS | Frederick Allan | 1st Bn |
| WARRANT OFFICERS CLASS II (CSM) | | | | | |
| 13 | 28.03.53 | SC41588 | DORAN, CD | Jean Paul | 1st Bn (also listed as 1st Bn PPCLI) |
| 40 | 06.10.51 | SD46009 | FULLER | George Hubert | 2nd Bn |
| 13 | 28.03.53 | SP16427 | GRAY | Kenneth Henry | 1st Bn |
| | | not listed | McNALLY, BEM | Richard | 1st Bn |
| SERGEANTS | | | | | |
| 15 | 12.04.52 | D800679 | DEE | Vincent | 2nd Bn |
| 7 | 01.06.53 | SK2915 | SCOOLAR | Thomas Raymond | 1st Bn (also listed as SCOULAR) |
| 45 | 08.11.52 | SB801126 | SILLABY | Harry Arthur | 2nd Bn |
| CORPORALS | | | | | |
| | | not listed | FAULKNER | Ellroy Morton | 1st Bn |
| | | not listed | GINGRAS | Paul Francis | 2nd Bn |
| | | not listed | McNULTY | Robert Ronald | 1st Bn |
| | | not listed | MORIN | Joseph Aime | 1st Bn |
| | | not listed | WHITNEY | Ronald Arthur | 1st Bn |
| LANCE-CORPORALS | | | | | |
| | | not listed | ALGEE | Burnell Gordon | 1st Bn |
| 45 | 08.11.52 | SB7236 | BOLTON | Russell Wilburn | 1st Bn (also listed as 2nd Regiment RCHA) |
| | | not listed | LUDGATE | Clarence Edward | 1st Bn |

| Canadian/ London Gazette issue No. | Canadian/ London Gazette date | Personal Service number | Name | Christian names | Remarks |
|---|---|---|---|---|---|
| MENTION IN DESPATCHES (FOR GALLANT AND DISTINGUISHED SERVICES) (continued) | | | | | |
| BRONZE OAK LEAF EMBLEM | | | | | |
| PRIVATES | | | | | |
| | | not listed | ARSENAULT | Earl Joseph | 1st Bn |
| 45 | 08.11.52 | C800054 | CALDWELL | Raymond | 2nd Bn |
| 45 | 08.11.52 | SC9124 | COX | Thomas Walter | 1st Bn |
| 45 | 08.11.52 | B801935 | FERGUSON | William Connor | 2nd Bn |
| 40 | 03.10.53 | SC136005 | JUNKINS | John Sinclair | 3rd Bn (also listed as 3rd Bn PPCLI) |
| 15 | 12.04.52 | B800429 | KANE | Frank Patrick | 2nd Bn |
| 15 | 12.04.52 | G800054 | McINTYRE | George Leonard | 2nd Bn (also list as McINTRY) |
| | | not listed | MORRISON | Charles Joseph | 1st Bn |
| 15 | 12.04.52 | B800206 | SHARPE | James Docherty | 2nd Bn |
| 40 | 06.10.51 | B800296 | TREMBLAY | John Daniel | 2nd Bn |
| 45 | 08.11.52 | A800005 | WINDLE | William Robert | 2nd Bn |

## AWARDS CONFERRED BY THE PRESIDENT OF THE UNITED STATES OF AMERICA

| Canadian/ London Gazette issue No. | Canadian/ London Gazette date | Personal Service number | Name | Christian names | Remarks |
|---|---|---|---|---|---|
| BRONZE STAR | | | | | |
| MAJOR | | | | | |
| | | not listed | MEDLAND, DSO | Richard Dillon | 2nd Bn |
| DISTINGUISHED FLYING CROSS | | | | | |
| LIEUTENANT | | | | | |
| | | not listed | MAGEE | Arthur Jeffrey | 1st Bn (attached 614th Tactical Control Group, 5th US Air Force) |

| London Gazette issue No. | London Gazette date | Personal Service number | Name | Christian names | Remarks |
|---|---|---|---|---|---|
| ROYAL 22[e] RÉGIMENT | | | | | |
| DSO: DISTINGUISHED SERVICE ORDER | | | | | |
| LIEUTENANT-COLONEL | | | | | |
| 40102 | 16.02.54 | ZP1337 | POULIN, CD | Jean Louis Gaston | Officer Commanding 3rd Bn |
| MAJOR | | | | | |
| 39467 | 12.02.52 | ZL300 | LIBOIRIN | Réal | 2nd Bn (also listed as LIBOIRON) |
| OBE: OFFICER OF THE ORDER OF THE BRITISH EMPIRE | | | | | |
| LIEUTENANT-COLONELS | | | | | |
| 39518 | 18.04.52 | ZD4622 | DEXTRAZE, DSO | Jaques Alfred | Officer Commanding 2nd Bn |
| 39819 | 07.04.53 | ZP1276 | TRUDEAU, DSO, CD | Louis Freemont | Officer Commanding 1st Bn |
| 39693 | 11.11.52 | ZP1345 | VALLEE, CD | Joseph Alexandre Armand Gaston | Officer Commanding 2nd Bn |
| MBE: MEMBER OF THE ORDER OF THE BRITISH EMPIRE | | | | | |
| MAJORS | | | | | |
| 39980 | 06.10.53 | ZE2158 | SEVIGNY, DSO | Joseph Georges | 1st Bn |
| 40102 | 16.02.54 | ZF1435 | TURCOTTE | Lucien René Peirre Gustave | 3rd Bn |
| CAPTAIN | | | | | |
| 39980 | 06.10.53 | ZC786 | DUBOIS | Albiny | 3rd Bn |
| WARRANT OFFICER CLASS I (RSM) | | | | | |
| 39980 | 06.10.53 | SE4001 | HACHE, CD | Patrick | 1st Bn |
| MC:MILITARY CROSS | | | | | |
| MAJOR | | | | | |
| 39885 | 12.06.53 | ZD220 | POPE | William Henry | 1st Bn |
| CAPTAINS | | | | | |
| 39693 | 11.11.52 | ZE337 | Le CLERC | Roland | 2nd Bn (also listed as LECLERC) |
| 39622 | 12.08.52 | ZD2770 | TREMBLAY | Joseph Patrick René | 2nd Bn |
| LIEUTENANT | | | | | |
| 39622 | 12.08.52 | ZD4529 | THERRIEN | Jean Paul André | 2nd Bn |
| 2ND LIEUTENANTS | | | | | |
| 39819 | 07.04.53 | ZG10026 | MERRITHEW | Haldene Owen | 1st Bn |
| 40102 | 16.02.54 | ZE10423 | RIFFOU | Jean Berchmans | 3rd Bn |
| BAR TO DCM: DISTINGUISHED CONDUCT MEDAL | | | | | |
| CORPORAL | | | | | |
| 39467 | 12.02.52 | 800999 | MAJOR, DCM | Leo | 2nd Bn |

| Canadian/ London Gazette issue No. | Canadian/ London Gazette date | Personal Service number | Name | Christian names | Remarks |
|---|---|---|---|---|---|
| **DCM: DISTINGUISHED CONDUCT MEDAL** | | | | | |
| **LANCE-CORPORAL** | | | | | |
| 39467 | 12.02.52 | D801889 | HARVEY | Joseph Paul André | 2nd Bn |
| **MM: MILITARY MEDAL** | | | | | |
| **SERGEANTS** | | | | | |
| 39622 | 12.08.52 | E800061 | BEAUDIN | Arthur | 2nd Bn |
| 39881 | 06.10.53 | SE103492 | BERGERON | Bruno | 1st Bn |
| 39819 | 07.04.53 | SD189946 | CHAMPOUX | Joseph Renaud | 1st Bn |
| 39433 | 04.01.52 | E800043 | SOMMERVILLE | Samuel | 2nd Bn |
| **CORPORALS** | | | | | |
| 39589 | 04.07.52 | SE800308 | CORMIER | Delphis | 2nd Bn |
| 39433 | 04.01.52 | D801131 | OSTIGUY | Jean Gérard | 2nd Bn |
| 39819 | 07.04.53 | SB7474 | PEARCE | Jean Paul Roland | 1st Bn |
| **LANCE-CORPORALS** | | | | | |
| 39819 | 07.04.53 | SE800476 | DION | Antoine | 1st Bn |
| 39819 | 07.04.53 | SC850324 | GINGRAS | Jean Robert | 1st Bn |
| **PRIVATES** | | | | | |
| 39518 | 18.04.52 | SE6576 | ASSELIN | Ernest | 2nd Bn |
| 39433 | 04.01.52 | SE800150 | GAGNON | Romeo | 2nd Bn |
| 39646 | 16.09.52 | D801793 | GUAY | Jean Guy | ← Unit not listed (also shown as L/CPL) |
| **BEM: BRITISH EMPIRE MEDAL** | | | | | |
| **(FOR GALLANT AND DISTINGUISHED SERVICES WHILST A PRISONER OF WAR IN NORTH KOREA)** | | | | | |
| **LANCE-CORPORAL** | | | | | |
| 40115 | 02.03.54 | SE6785 | DUGAL | Paul | 1st Bn |
| **BEM: BRITISH EMPIRE MEDAL** | | | | | |
| **SERGEANTS** | | | | | |
| 39885 | 01.06.53 | SE103636 | BORDEAU | Jean | 1st Bn |
| 40102 | 12.02.54 | SD190988 | CHARLAND | Bernard Irenee | 3rd Bn |
| 39980 | 06.10.53 | SD81578 | STEWART | Charles André | 3rd Bn |
| **MENTION IN DESPATCHES (FOR GALLANT AND DISTINGUISHED SERVICES - POSTHUMOUS)** | | | | | |
| **BRONZE OAK LEAF EMBLEM** | | | | | |
| **CORPORAL** | | | | | |
| 40 | 03.10.53 | SE103594 | MORIN | Joseph Aime | 1st Bn |
| **PRIVATE** | | | | | |
| 45 | 08.11.52 | D801844 | DUPOIS | Joseph Elphege Albert | 1st Bn |

| Canadian/ London Gazette issue No. | Canadian/ London Gazette date | Personal Service number | Name | Christian names | Remarks |
|---|---|---|---|---|---|
| MENTION IN DESPATCHES (FOR GALLANT AND DISTINGUISHED SERVICES) | | | | | |
| BRONZE OAK LEAF EMBLEM | | | | | |
| MAJORS | | | | | |
| 7 | 13.02.54 | ZE4127 | BERNIER | Yvan | 3rd Bn |
| 40 | 06.10.52 | ZE1110 | DUBE | Joseph Leopold Yvan | 2nd Bn (listed: deceased) |
| 13 | 28.03.53 | ZC2200 | PELLETIER | Joseph Ernest Modiste Bernard | 1st Bn (initials also listed as JAMB) |
| 45 | 08.11.52 | ZE219 | PERUSSE | Joseph Patrick Gabriel | 2nd Bn |
| 45 | 08.11.52 | ZD365 | SUTHERLAND | William Edward | 2nd Bn |
| CAPTAINS | | | | | |
| 5 | 31.01.53 | ZD4610 | BOUFFARD | Paul Jacques Clement | 2nd Bn (initials also listed as JPJC) |
| 15 | 12.04.52 | ZE3307 | FORBES | Jean Charles Bertrand | 2nd Bn |
| LIEUTENANTS | | | | | |
| 7 | 13.02.54 | ZC9533 | BOWEN | Gerald Reidy | 3rd Bn |
| 15 | 12.04.52 | TE65298 | BROUARD | Camille | 2nd Bn |
| 40 | 03.10.53 | ZD9930 | FOURNIER | Jean Marc | 1st Bn |
| | | not listed | McDUFF | Raymond | 2nd Bn |
| | | not listed | NASH | Walter George Joseph | 2nd Bn |
| | | not listed | PARADIS | Jacques | 1st Bn (also listed as 2nd Bn) |
| 40 | 06.10.51 | ZE4132 | LOUFFE | Joseph Fanida Orphila | 2nd Bn |
| 2ND LIEUTENANTS | | | | | |
| 7 | 13.02.54 | ZE9620 | BOITEAU | Joseph Paul Denis | 3rd Bn |
| 7 | 13.02.54 | ZB10243 | CAMPBELL | William Roy | 3rd Bn |
| 45 | 08.11.52 | ZE10135 | FOURNIER | Real | 1st Bn |
| 7 | 01.06.53 | ZL10018 | FORNESS | Frank Paul | 1st Bn |
| WARRANT OFFICER CLASS I (RSM) | | | | | |
| 5 | 31.01.53 | SE4301 | DAGEMAIS | George | 2nd Bn (also listed as DAGENAIS) |
| WARRANT OFFICER CLASS II (CSM) | | | | | |
| 7 | 01.06.53 | SD61013 | DUSSAULT | Hermegilde | 1st Bn |
| STAFF SERGEANT | | | | | |
| 13 | 28.03.53 | SD41123 | VALLIERE | Charles Henri | 1st Bn (also listed as VALLIERES) |
| SERGEANTS | | | | | |
| 13 | 28.03.53 | SD190368 | BROWN | John Aubrey | 1st Bn |
| 15 | 12.04.52 | D800693 | ZALUSKI | Walter | 2nd Bn |
| CORPORALS | | | | | |
| 15 | 12.04.52 | D800676 | CHARPENTIER | March | 2nd Bn (listed: deceased) |
| | | not listed | GINGRAS | Paul Francis | 2nd Bn |
| | | not listed | ISTEAD | Earl, Jnr | 2nd Bn |
| 47 | 21.11.53 | SD4552 | LADOUCEUR | Joseph Gaston Maurice | 1st Bn (listed: deceased) |

| Canadian/ London Gazette issue No. | Canadian/ London Gazette date | Personal Service number | Name | Christian names | Remarks |
|---|---|---|---|---|---|
| MENTION IN DESPATCHES (FOR GALLANT AND DISTINGUISHED SERVICES) (continued) | | | | | |
| BRONZE OAK LEAF EMBLEM | | | | | |
| CORPORALS (continued) | | | | | |
| 7 | 13.02.54 | SE6499 | LANGLOIS | Eugene | 3rd Bn |
| | | not listed | PRUD'HOMME | Daniel | 2nd Bn |
| 13 | 28.03.53 | SD4512 | RIOUX | Joseph Andre | 1st Bn |
| LANCE-CORPORALS | | | | | |
| | | SC850324 | GINGRAS, MM | Jean Robert | 1st Bn |
| 40 | 03.10.53 | SE6727 | PAQUETTE | Jean | 3rd Bn (also listed as PAQUET) |
| PRIVATES | | | | | |
| 15 | 12.04.52 | G800368 | DOUCETTE | Cornelius Joseph | 2nd Bn |
| 45 | 08.11.52 | E800381 | DUBOIS | Real | 2nd Bn |
| 31 | 02.08.52 | B802044 | GAUTHIER, MM | Andrew Joseph | 2nd Bn |
| 15 | 12.04.52 | SD802313 | MAHONEY | Brendon Thomas | 2nd Bn |
| 40 | 03.10.53 | SE6327 | PELLETIER | Alphones | 1st Bn |

AWARDS CONFERRED BY THE PRESIDENT OF THE UNITED STATES OF AMERICA

| | | | | | |
|---|---|---|---|---|---|
| DISTINGUISHED FLYING CROSS | | | | | |
| CAPTAIN | | | | | |
| | | not listed | YELLE | Joseph Roland Peirre Paul | 1st Bn (att US Air Force) |
| LIEUTENANT | | | | | |
| | | ZE4132 | PLOUFFE | Joseph Fanida Orphila | 2nd Bn |
| AIR MEDAL | | | | | |
| CAPTAIN | | | | | |
| | | not listed | DRAPEAU, DCM | Louis René | 2nd Bn |

BELGIAN AWARDS

| | | | | | |
|---|---|---|---|---|---|
| CHEVALIER DE L'ORDRE DE LÉOPOLD II AVEC PALME AND LA CROIX DE GUERRE 1940 AVEC PALME | | | | | |
| LIEUTENANT | | | | | |
| | | not listed | GAGNE | Jean | 3rd Bn |
| DECORATION MILITAIRE 2[e] CLASSE AVEC PALME AND LA CROIX DE GUERRE 1940 AVEC PALME | | | | | |
| CORPORAL | | | | | |
| | | not listed | PORTELANCE | Roger | 3rd Bn |

| London Gazette issue No. | London Gazette date | Personal Service number | Name | Christian names | Remarks |
|---|---|---|---|---|---|
| PRINCESS PATRICIA'S CANADIAN LIGHT INFANTRY | | | | | |
| 2ND BAR TO DSO: DISTINGUISHED SERVICE ORDER | | | | | |
| LIEUTENANT-COLONEL | | | | | |
| 39518 | 18.04.52 | ZK4311 | STONE, DSO, MC | James Riley | Officer Commanding 2nd Bn 27th., 28th. and 25th Brigades |
| DSO: DISTINGUISHED SERVICE ORDER | | | | | |
| LIEUTENANT-COLONEL | | | | | |
| 39693 | 11.11.52 | ZH1136 | WILSON-SMITH, MBE | Norman George | Officer Commanding 1st Bn |
| MAJOR | | | | | |
| 39494 | 18.03.52 | ZM4637 | GEORGE | John Herbert Bothwell | 1st Bn |
| OBE: OFFICER OF THE ORDER OF THE BRITISH EMPIRE | | | | | |
| LIEUTENANT-COLONELS | | | | | |
| 39819 | 07.04.53 | ZF276 | CAMERON | John Ralph | Officer Commanding 1st Bn |
| 40102 | 16.02.54 | ZF2555 | MacLACHLAN, MC, CD | Malcolm Francis | Officer Commanding 3rd Bn (also listed as McLACHLAN) |
| MBE: MEMBER OF THE ORDER OF THE BRITISH EMPIRE | | | | | |
| MAJORS | | | | | |
| 39693 | 11.11.52 | ZM102 | ALLAN, DSO, CD | James Chalmers | att HQ 25 Infantry Brigade |
| 39885 | 12.06.53 | ZH738 | BRUCE, CD | Robert Fraser | 3rd Bn |
| 39980 | 06.10.53 | ZK122 | MacNEILL, ED | Charles Edward Collie | ← Unit not listed |
| LIEUTENANT | | | | | |
| 39819 | 07.04.53 | ZM4100 | WOOD | Michael Bruce | 1st Bn |
| MC: MILITARY CROSS | | | | | |
| CAPTAIN | | | | | |
| 39282 | 10.07.5 | ZH4018 | MILLS | John Graham Wallace | 2nd Bn |
| LIEUTENANTS | | | | | |
| 39518 | 10.05.52 | ZK5089 | McKINLEY | James George Clyde | 1st Bn |
| 39646 | 16.09.52 | ZH9926 | MIDDLETON, MM | David Alexander | 1st Bn |
| 39981 | 06.10.52 | ZK4698 | PITTS | Herbert Chesley | 3rd Bn |
| 2ND LIEUTENANTS | | | | | |
| 39819 | 07.04.53 | ZL9929 | ROBERTSON | William Cryle | 1st Bn |
| 40102 | 16.02.54 | ZB10256 | SNIDER | Christopher Burnet | 3rd Bn |

| Canadian/ London Gazette issue No. | Canadian/ London Gazette date | Personal Service number | Name | Christian names | Remarks |
|---|---|---|---|---|---|
| DCM: DISTINGUISHED CONDUCT MEDAL | | | | | |
| SERGEANTS | | | | | |
| 39601 | 18.07.52 | SX3460 | BUXTON | Richard George | 1st Bn |
| 39433 | 04.01.52 | H800021 | McCUISH | David Allan | 2nd Bn |
| 39759 | 23.01.53 | SK12487 | RICHARDSON | John Henry | 1st Bn |
| PRIVATE | | | | | |
| 39354 | 19.10.51 | M800148 | MITCHELL | Wayne Robert | 2nd Bn |
| MM: MILITARY MEDAL | | | | | |
| STAFF SERGEANT | | | | | |
| 39981 | 06.10.53 | SB90023 | COLE | Vernon David | 3rd Bn |
| SERGEANTS | | | | | |
| 39433 | 04.01.52 | H800926 | DUNPHY | Kerry John | 2nd Bn |
| 39819 | 07.04.53 | SB153746 | PRENTICE | Rhodes Albert | 1st Bn |
| 39981 | 06.10.53 | SB167977 | STEADMAN | Leonard | 3rd Bn (also listed as Cpl in Canadian official listings) |
| CORPORALS | | | | | |
| 39646 | 16.09.52 | SK8338 | DUNBAR | John Glenford | 1st Bn |
| 39819 | 07.04.53 | SK14630 | FENTON | Vincent Lloyd | 1st Bn |
| 39369 | 11.11.52 | SL918 | RIMMER | James Ernest | 1st Bn |
| LANCE-CORPORALS | | | | | |
| 39282 | 10.07.52 | M800022 | DOUGLAS | Smiley | 2nd Bn |
| 40102 | 16.02.54 | SB12691 | THOMPSON | Phillip Charles | 3rd Bn |
| PRIVATES | | | | | |
| 39282 | 10.07.52 | B801304 | BARTON | Leonard | 2nd Bn |
| 39467 | 12.02.52 | M800036 | BARWISE | Kenneth Francis | 2nd Bn |
| BEM: BRITISH EMPIRE MEDAL | | | | | |
| WARRANT OFFICER CLASS II (CSM) | | | | | |
| 39819 | 07.04.53 | SD82512 | LYNCH | Paul Austin | 1st Bn |
| SERGEANT | | | | | |
| 39693 | 11.11.52 | SK14437 | ELLIOTT | George William | 1st Bn (also listed as ELLIOT) |
| MENTION IN DESPATCHES (FOR GALLANT AND DISTINGUISHED SERVICES) | | | | | |
| BRONZE OAK LEAF EMBLEM | | | | | |
| MAJORS | | | | | |
| 45 | 08.11.52 | ZM1099 | SHORT, MC | Charles Gilmour | 1st Bn |
| 40 | 03.10.53 | ZL849 | WILSON, CD | Walter Maurice Woodrow | 3rd Bn |

| Canadian/ London Gazette issue No. | Canadian/ London Gazette date | Personal Service number | Name | Christian names | Remarks |
|---|---|---|---|---|---|
| MENTION IN DESPATCHES (FOR GALLANT AND DISTINGUISHED SERVICES) (continued) | | | | | |
| BRONZE OAK LEAF EMBLEM | | | | | |
| CAPTAINS | | | | | |
| 13 | 28.03.53 | ZK794 | BERGER | Torleif | 1st Bn |
| 15 | 12.04.54 | ZD761 | GUNTON | Gordon Arthur | 1st Bn |
| 7 | 13.02.54 | ZM427 | KEMSLEY | Charles Arthur | 3rd Bn (also listed as KENSLEY) |
| LIEUTENANTS | | | | | |
| 13 | 28.03.53 | ZM9776 | BELL | Marks Robert | 1st Bn |
| 15 | 12.04.52 | ZH3975 | DAVIS | John MacArthur | 1st Bn |
| 45 | 08.11.52 | ZD4097 | HARTNETT | James Joseph | 2nd Bn (also listed as 2nd Bn RCR) |
| 40 | 06.10.51 | TB1794 | WEBB | Thomas Robert | 2nd Bn |
| WARRANT OFFICERS CLASS II (CSM) | | | | | |
| 13 | 28.03.53 | SP28012 | CLIFTON | Ashworth Thomas | 1st Bn |
| 13 | 28.03.53 | SC41588 | DORAN, CD | Jean Paul | 1st Bn (also listed as 1st Bn RCR) |
| STAFF SERGEANT | | | | | |
| 13 | 28.03.53 | SH45752 | APPLETON, MM | Roy Harold | 1st Bn |
| SERGEANTS | | | | | |
| 45 | 08.11.52 | SB801995 | LOCK | William Alfred | 1st Bn |
| 15 | 12.04.52 | SL111324 | STONE | Joseph Ernest | 1st Bn |
| CORPORALS | | | | | |
| 45 | 08.11.52 | SH61543 | McCLEOD | George Stanley | 1st Bn (initials also shown as D.S.) |
| 13 | 28.03.53 | SL108617 | LIPPUS | Vernon Ray | 1st Bn |
| 13 | 28.03.53 | SK14497 | REDING | Raymond George | 1st Bn |
| LANCE-CORPORALS | | | | | |
| 7 | 01.06.53 | SK13736 | FREUND | Peter | 3rd Bn |
| 15 | 12.04.52 | K800043 | HOFFMAN | Cyrille Cedric | 2nd Bn |
| 13 | 18.03.53 | SK9545 | INKMAN | Ernest Peter | 1st Bn |
| 15 | 12.04.52 | SM9332 | MATHEWS | Glen Abbott | 1st Bn |
| 13 | 28.03.53 | SC8728 | MURPHY | Pierce Leo | 1st Bn |
| 45 | 08.11.52 | SM9228 | ROONEY | Donald Hubert | 1st Bn |
| PRIVATES | | | | | |
| 13 | 28.03.53 | SA1502 | BERTRAND | Wilfred Laurier | 1st Bn |
| 15 | 12.04.52 | B801350 | COPLEY | Donald R. | 2nd Bn |
| 7 | 13.02.54 | SB10854 | HUTCHENS | Carson James Joseph | 3rd Bn |
| 7 | 13.02.54 | SK13805 | JOHNSTON | Norris Hyde | 3rd Bn |
| 40 | 03.10.53 | SC136005 | JUNKINS | John Sinclair | 3rd Bn (also listed as 3rd Bn RCR) |
| | | not listed | KAWANAMI | Masao | 2nd Bn |
| 15 | 12.04.52 | SM800357 | RAGAN | Harold Richard | 2nd Bn |
| 40 | 03.10.53 | SL4798 | SCHOFIELD | Murray | 3rd Bn |
| 7 | 13.02.54 | SH704 | SHALER | Glen David | 3rd Bn |
| 13 | 28.03.53 | SD4852 | WOODS | Russell Larry | 1st Bn |

| Canadian/ London Gazette issue No. | Canadian/ London Gazette date | Personal Service number | Name | Christian names | Remarks |
|---|---|---|---|---|---|

AWARDS CONFERRED BY THE PRESIDENT OF THE UNITED STATES OF AMERICA

BRONZE STAR WITH "V" DEVICE
MAJOR

| | | | | | |
|---|---|---|---|---|---|
| | | not listed | WILLIAMS | Edward Geffery | 1st Bn |

BRONZE STAR
MAJOR

| | | | | | |
|---|---|---|---|---|---|
| | | not listed | HUGGARD | Charles Otis | 1st Bn |

AIR MEDAL
LIEUTENANTS

| | | | | | |
|---|---|---|---|---|---|
| | | not listed | BULL | Albert Paul | 1st Bn |
| | | not listed | MacLEOD | Donald Gordon | 2nd Bn |

2ND LIEUTENANT

| | | | | | |
|---|---|---|---|---|---|
| | | ZL9929 | ROBERTSON | William Cryle | 1st Bn |

BELGIAN AWARD

OFFICER DE L'ORDRE DE LÉOPOLD II AVEC PALME AND LA CROIX DE GUERRE 1940 AVEC PALME

| | | | | | |
|---|---|---|---|---|---|
| CAPTAIN | | not listed | MARCHESSAULT, CD | Marcel Henri | 3rd Bn |

| Canadian/ London Gazette issue No. | Canadian/ London Gazette date | Person Service number | Name | Christian names | Remarks |
|---|---|---|---|---|---|
| ROYAL CANADIAN ARMY CHAPLAINS' DEPARTMENT | | | | | |
| MBE: MEMBER OF THE ORDER OF THE BRITISH EMPIRE | | | | | |
| CAPTAINS | | | | | |
| 39819 | 07.04.53 | ZB4355 | REVEREND FILSHIE | James Alexander | att 1st & 2nd Bn PPCLI |
| 39980 | 06.10.53 | ZE4586 | REVEREND FORTIN | Gerard Raymond | att 2nd Bn Royal 22e Regiment and HQ 25 Canadian Infantry Brigade |
| 39980 | 06.10.53 | ZA4711 | REVEREND JOHNSON | Howard Wesley | att 25 Canadian Field Ambulance |
| 39885 | 09.06.53 | ZF4289 | REVEREND MacGREGOR | John Hector | att 2nd Bn PPCLI |
| MENTION IN DESPATCHES (FOR GALLANT AND DISTINGUISHED SERVICES) | | | | | |
| BRONZE OAK LEAF EMBLEM | | | | | |
| MAJORS | | | | | |
| 40 | 03.10.53 | ZD4099 | REVEREND BEAUPARLANT | Joseph Ronault | att RCASC 25 Canadian Infantry Brigade |
| 40 | 03.10.53 | ZG3152 | REVEREND FRASER | Gordon Sutherland | att HQ 25 Canadian Infantry Brigade |
| CAPTAINS | | | | | |
| 45 | 08.11.52 | ZE4728 | REVEREND LEBEL | Contram | att 2nd Bn Royal 22e Regiment |
| 13 | 28.03.53 | ZD4044 | REVEREND PATENAUDE | Joseph Arthur Oliver | att 1st Bn Royal 22e Regiment |

| Canadian/ London Gazette issue No. | Canadian/ London Gazette date | Personal Service number | Name | Christian names | Remarks |
|---|---|---|---|---|---|
| ROYAL CANADIAN ARMY SERVICE CORPS | | | | | |
| MBE: MEMBER OF THE ORDER OF THE BRITISH EMPIRE | | | | | |
| MAJORS | | | | | |
| 39693 | 11.11.52 | ZB2117 | BRODIE | Harry Band | HQ 1st Commonwealth Division |
| 39819 | 07.04.53 | ZA483 | DOLAN | Joseph Ignatius | 23 Transport Company |
| 40102 | 16.02.54 | ZP1147 | HESSION | Edmond Gilbert | 56 Transport Company |
| 39622 | 12.08.52 | ZC2944 | LAUGHTON | Robert Charles David | 54 Transport Company |
| MM: MILITARY MEDAL | | | | | |
| PRIVATE | | | | | |
| 39433 | 04.01.52 | G800518 | WHITE | Ottie Malcolm | att 25 Canadian Field Ambulance |
| BEM: BRITISH EMPIRE MEDAL | | | | | |
| STAFF SERGEANT | | | | | |
| | | not listed | CLOUSTON | Byron Clifford | Canadian Military Mission Far East |
| SERGEANT | | | | | |
| 39693 | 11.11.52 | SH800192 | HAYNES | Malcolm Stuart | Canadian Section of Base Troops, BCFK |
| MENTION IN DESPATCHES (FOR GALLANT AND DISTINGUISHED SERVICES) | | | | | |
| BRONZE OAK LEAF EMBLEM | | | | | |
| MAJORS | | | | | |
| 40 | 03.10.53 | ZP1651 | HARPER, MBE, CD | Dennis Alfred | HQ 25 Canadian Infantry Brigade |
| 13 | 28.03.53 | ZC470 | ROCHEFORT, MBE | Joseph Arthur René | HQ 1st Commonwealth Division |
| 31 | 02.08.52 | ZB2724 | WHITICAR, CD | Clarence Mervyn | ← Unit not listed |
| CAPTAINS | | | | | |
| 15 | 12.04.52 | ZM2911 | DAY | Richard MacKreth | 54 Transport Company |
| 13 | 28.03.53 | ZP4753 | FISHER, CD | Orland Howatt | 23 Transport Company |
| 40 | 03.10.53 | ZF851 | LEONARD, CD | Lloyd | HQ 25 Canadian Infantry Brigade |
| 7 | 01.06.53 | ZB854 | MAYES | Kenneth Melvin | 23 Transport Company |
| 15 | 12.04.52 | ZB2915 | McKENZIE | John Henry | HQ 25 Canadian Infantry Brigade |
| 40 | 03.10.53 | ZD182 | PELLAND | Joseph Louis Adrien Paul Tremblay | 23 Transport Company |
| 7 | 13.02.54 | ZB3022 | SMITH | George Roderick | 56 Transport Company |
| 7 | 01.06.53 | ZD2132 | TURTON | Charles William | HQ 25 Canadian Infantry Brigade |
| 45 | 08.11.52 | ZC4072 | URGUART | William Robert | 54 Transport Company (also listed URQUHART) |
| 7 | 13.02.54 | ZP2390 | WARD, CD | Earl Caldwell | att 38 Canadian Field Ambulance |

| Canadian/ London Gazette issue No. | Canadian/ London Gazette date | Personal Service number | Name | Christian names | Remarks |
|---|---|---|---|---|---|
| MENTION IN DESPATCHES (FOR GALLANT AND DISTINGUISHED SERVICES)(continued) | | | | | |
| BRONZE OAK LEAF EMBLEM | | | | | |
| WARRANT OFFICER CLASS I | | | | | |
| 13 | 28.03.53 | not listed | GIBBS, CD | Walter Robert | 25 Transport Company (also listed 23 Transport Company) |
| STAFF SERGEANT | | | | | |
| 40 | 03.10.53 | SC58804 | LEPAGE | Earl Vincent | HQ 1st Commonwealth Division (also listed with initial R. only) |
| SERGEANTS | | | | | |
| 40 | 06.10.51 | F800140 | CARSON | James Cuthbert | * Unit not listed |
| 40 | 03.10.53 | SA21450 | CHEESEMAN | John Ross | 56 Transport Company |
| 15 | 12.04.52 | B801063 | CHILLMAN | John William | 54 Transport Company |
| 15 | 12.04.52 | SB60804 | HUNT, MM | George Mervyn | att 27 Canadian Field Ambulance (also listed as RCAMC) |
| CORPORAL | | | | | |
| 40 | 03.10.53 | SA27140 | BRINE | Louis Ford | HQ 25 Canadian Infantry Brigade |
| PRIVATES | | | | | |
| 40 | 03.10.53 | SH4791 | BREEN | Harold Frederick | att 38 Canadian Field Ambulance (redesignated No.4 Field Ambulance) |
| 31 | 02.08.52 | H800412 | CLOUTIER | Joseph John | 54 Transport Company |
| 40 | 03.10.53 | SB12513 | ESSEX | Richard Neale | HQ 1st Commonwealth Division |
| 40 | 03.10.53 | SB7291 | FISHER | Harold James | HQ 1st Commonwealth Division |
| 15 | 12.04.52 | SN1111 | TOBIN | Alexander Joseph | att 38 Canadian Field Ambulance (redesignated No.4 Field Ambulance December, 1953) |

| London Gazette issue No. | London Gazette date | Personal Service number | Name | Christian names | Remarks |
|---|---|---|---|---|---|
| ROYAL CANADIAN ARMY MEDICAL CORPS | | | | | |
| CBE: COMMANDER OF THE ORDER OF THE BRITISH EMPIRE | | | | | |
| COLONEL | | | | | |
| 39885 | 12.06.53 | ZP1438 | SMITH, OBE, CD | Gerald Morgan | Senior Medical Officer HQ 1st Commonwealth Division |
| OBE: OFFICER OF THE ORDER OF THE BRITISH EMPIRE | | | | | |
| LIEUTENANT-COLONELS | | | | | |
| 39622 | 12.08.52 | ZB553 | BROSSEAU, MC | Bernard Louis Persillier | *25 Canadian Field Ambulance |
| 39819 | 07.04.53 | ZB2391 | CASWELL, MC | Clive Browning | *37 Canadian Field Ambulance |
| 40102 | 16.02.54 | ZA3000 | GALLOWAY | James Duncan | *38 Canadian Field Ambulance (redesignated No.4 Field Ambulance, December, 1953) |
| MBE: MEMBER OF THE ORDER OF THE BRITISH EMPIRE | | | | | |
| LIEUTENANT-COLONEL | | | | | |
| 39885 | 12.06.53 | ZB2944 | SMILLIE | Roy Alexander | *38 Canadian Field Ambulance (redesignated No.4 Field Ambulance) |
| MAJORS | | | | | |
| 39693 | 11.11.52 | ZC2858 | HITSMAN | James Stuart | *25 Canadian Field Ambulance |
| 39819 | 07.04.53 | ZB2317 | JAFFEY | Bertram David | 37 Canadian Field Ambulance |
| CAPTAIN | | | | | |
| 39518 | 18.04.52 | TH9140 | STEVENSON | Henry Carson | att 2nd Bn Royal Canadian Regiment |
| DCM: DISTINGUISHED CONDUCT MEDAL | | | | | |
| CORPORAL | | | | | |
| 39518 | 18.04.52 | L800192 | POOLE | Ernest William | att 2nd Bn Royal Canadian Regiment |
| GM: GEORGE MEDAL | | | | | |
| LANCE-CORPORAL | | | | | |
| 39604 | 22.07.52 | SG2594 | SINNOTT | Sterling Lloyd | 25 Canadian Field Ambulance |
| MM: MILITARY MEDAL | | | | | |
| CORPORAL | | | | | |
| | | SH61702 | McKINNEY | Gerald Allen | att 1st Bn Royal Canadian Regiment |
| BEM: BRITISH EMPIRE MEDAL | | | | | |
| SERGEANT | | | | | |
| | 10.04.52 | SF800173 | PARKER | James Willard | 25 Canadian Field Ambulance |

* Officers Commanding

| Canadian/ London Gazette issue No. | Canadian/ London Gazette date | Personal Service number | Name | Christian names | Remarks |
|---|---|---|---|---|---|
| BEM: BRITISH EMPIRE MEDAL (continued) | | | | | |
| CORPORAL | | | | | |
| 39980 | 06.10.53 | SB33432 | DOWNS | William Malcolm | att 3rd Bn Royal Canadian Regiment |
| MENTION IN DESPATCHES (FOR GALLANT AND DISTINGUISHED SERVICES) | | | | | |
| BRONZE OAK LEAF EMBLEM | | | | | |
| MAJORS | | | | | |
| 45 | 08.11.52 | ZD4752 | ROBITAILLE | Robert Joseph Albert | HQ 1st Commonwealth Division |
| 40 | 03.10.53 | ZF1924 | WATSON | James Hubert | 37 Canadian Field Ambulance |
| CAPTAINS | | | | | |
| 31 | 02.08.52 | ZB3486 | BESLEY | John Keith | ← Unit not listed |
| 40 | 03.10.53 | ZP1592 | MUGGERIDGE, CD | Gordon Edward | 37 Canadian Field Ambulance |
| STAFF SERGEANT | | | | | |
| 40 | 03.10.53 | SB160969 | SCHAUBEL | Earl Vincent | 38 Canadian Field Ambulance (redesignated No.4 Field Ambulance) |
| SERGEANTS | | | | | |
| 15 | 12.04.52 | SB60804 | HUNT, MM | George Mervyn | 25 Canadian Field Ambulance |
| | | not listed | TAYLOR | Franck Willian Henry | att 2nd Bn PPCLI |
| CORPORALS | | | | | |
| 31 | 02.08.52 | B800357 | BLANCHIELD | Nelson Joseph | 25 Canadian Field Ambulance |
| | | not listed | BRAYTON | Ronald Joseph | att 3rd Bn Royal Canadian Regiment |
| 45 | 08.11.52 | SM52894 | CHESLAK | Michael Lawrence | 25 Canadian Field Ambulance |
| 45 | 08.11.52 | SF800752 | COOK | George Edgar | att 2nd Regiment RCHA |
| 13 | 28.03.53 | SF36233 | LEBLANC | Joseph Edgar | att 1st Bn R22$^{e}$ Regiment |
| 15 | 12.04.52 | SE103717 | MICHAUD | Jules | att 1st Bn PPCLI |
| 7 | 01.06.53 | SB7705 | NEWTON | William Roy | att 3rd Bn PPCLI |
| ROYAL RED CROSS CLASS I | | | | | |
| CAPTAIN (MATRON) | | | | | |
| 39885 | 12.06.53 | NP1289 | PENSE, ARRC, CD | Elizabeth Barker | Canadian Section: British Commonwealth General Hospital |
| ROYAL RED CROSS CLASS II | | | | | |
| LIEUTENANT (NURSING SISTER) | | | | | |
| 39981 | 06.10.53 | NH2891 | MacDONALD | Josephine Isabel | Canadian Section: British Commonwealth General Hospital |

AWARD CONFERRED BY THE PRESIDENT OF THE UNITED STATES OF AMERICA

LEGION OF MERIT, DEGREE OF LEGIONNAIRE

COLONEL

| | | | | | |
|---|---|---|---|---|---|
| | | ZP1438 | SMITH, CBE, CD | Gerald Lucien Morgan | HQ 1st Commonwealth Division |

| Canadian/ London Gazette issue No. | Canadian/ London Gazette date | Personal Service number | Name | Christian names | Remarks |
|---|---|---|---|---|---|
| ROYAL CANADIAN ARMY ORDNANCE CORPS | | | | | |
| MBE: MEMBER OF THE ORDER OF THE BRITISH EMPIRE | | | | | |
| MAJORS | | | | | |
| 39622 | 12.08.52 | ZG1128 | FERRIS | Howard Rowley | 25 Canadian Infantry Brigade Ordnance Company and 1st Commonwealth Division Ordnance Field Park |
| 40102 | 16.02.54 | ZC2548 | PRESTON | William Ross | 25 Canadian Infantry Brigade Ordnance Company and 1st Commonwealth Division Ordnance Field Park |
| CAPTAINS | | | | | |
| 39980 | 06.10.53 | ZC705 | BLAKE, CD | George Stewart | 191 Infantry Workshop CREME |
| 39980 | 06.10.53 | ZA2679 | CAMPBELL | Stanley Leonard | HQ 25 Canadian Infantry Brigade (also listed as CAMBELL) |
| MENTION IN DESPATCHES (FOR GALLANT AND DISTINGUISHED SERVICES) | | | | | |
| BRONZE OAK LEAF EMBLEM | | | | | |
| MAJOR | | | | | |
| | | ZG1128 | FERRIS | Harold Rowley | HQ 25 Canadian Infantry Brigade Ordnace Company and 1st Commonwealth Division Ordnance Field Park |
| SERGEANT | | | | | |
| 31 | 02.08.52 | SF35284 | MURRAY | Robert Fraser | ← Unit not listed |
| CORPORAL | | | | | |
| 40 | 06.10.51 | SC12678 | McEWAN | Gerald Edward | ← Unit not listed (name also listed as McELWAN) |
| PRIVATE | | | | | |
| 40 | 03.10.53 | SB77907 | MARSHALL | Eldon James | 1st Commonwealth Division Ordnance Field Park |

| Canadian/ London Gazette issue No. | Canadian/ London Gazette date | Personal Service number | Name | Christian names | Remarks |
|---|---|---|---|---|---|
| **ROYAL CANADIAN ARMY CORPS OF ELECTRICAL AND MECHANICAL ENGINEERS** | | | | | |
| **MBE: MEMBER OF THE ORDER OF THE BRITISH EMPIRE** | | | | | |
| MAJORS | | | | | |
| 39693 | 11.11.52 | ZA143 | McLARNON | John Robert | CREME HQ 1st Commonwealth Division |
| 39885 | 12.06.53 | ZD2016 | McLAUGHLIN | Ivan Murray | CREME HQ 1st Commonwealth Division |
| CAPTAINS | | | | | |
| 39693 | 11.11.52 | ZB2050 | McLAUGHLIN | Harold Edgar | 191 Infantry Workshops CREME |
| 39693 | 11.11.52 | ZB1363 | TROWER | Norman George | ← Unit not listed |
| LIEUTENANT | | | | | |
| 39805 | 09.06.53 | ZL3558 | LEONARD | Albert Clark | att B Squadron Lord Strathcona's Horse |
| WARRANT OFFICER CLASS I (RSM) | | | | | |
| | | not listed | FERNETS | J.M. | 191 Infantry Workshops CREME |
| **MENTION IN DESPATCHES (FOR GALLANT AND DISTINGUISHED SERVICES)** | | | | | |
| **BRONZE OAK LEAF EMBLEM** | | | | | |
| MAJOR | | | | | |
| 15 | 12.04.52 | ZK4587 | HALLAM | Richard Edwin | 191 Infantry Workshops CREME |
| CAPTAINS | | | | | |
| 40 | 03.10.53 | ZG2644 | FENDICK | Reginald Frost | 25 Field Workshops CREME (also listed as 23 Infantry Workshops) |
| 40 | 03.10.53 | ZK4201 | McLAUGHLIN | Peter Michael | HQ 25 Canadian Infantry Brigade (also listed as Infantry but Regiment not shown) |
| WARRANT OFFICER CLASS II | | | | | |
| 7 | 13.02.54 | SF93474 | COUGHRAN, CD | George Edward | 25 Field Workshops CREME (also listed as 23 Infantry Workshops) |
| STAFF SERGEANTS | | | | | |
| 7 | 01.06.53 | SF76750 | HICKS | Aubrey Layton | 191 Infantry Workshops CREME |
| 40 | 03.10.53 | SM44013 | McBLANE, CD | Alexander Gordon | 191 Infantry Workshops CREME |

| Canadian/ London Gazette issue No. | Canadian London Gazette date | Personal Service number | Name | Christian names | Remarks |
|---|---|---|---|---|---|
| ROYAL CANADIAN ARMY PAY CORPS | | | | | |
| MENTION IN DESPATCHES (FOR GALLANT AND DISTINGUISHED SERVICES) | | | | | |
| CAPTAIN | | | | | |
| 40 | 03.10.53 | ZC3085 | STEEL | Elmer Allison | HQ 25 Canadian Infantry Brigade |

| Canadian/ London Gazette issue No. | Canadian/ London Gazette date | Personal Service number | Name | Christian names | Remarks |
|---|---|---|---|---|---|
| CANADIAN MILITARY PROVOST CORPS | | | | | |
| MBE: MEMBER OF THE ORDER OF THE BRITISH EMPIRE | | | | | |
| MAJOR | | | | | |
| 39980 | 06.10.53 | ZM485 | LAWSON, CD | Quentin Earl | 25 Provost Detachment and 1st Comwel Division Provost Company |
| MENTION IN DESPATCHES (FOR GALLANT AND DISTINGUISHED SERVICES) | | | | | |
| BRONZE OAK LEAF EMBLEM | | | | | |
| LIEUTENANT | | | | | |
| 45 | 08.11.52 | ZK3412 | MURPHY | Robert Harold | 25 Field Punishment Camp and 25 Field Detention Barracks |
| WARRANT OFFICER CLASS II | | | | | |
| 40 | 03.10.53 | SA56040 | WOOD, CD | Leslie James | HQ 25 Infantry Brigade and 25 Field Detention Barracks |
| CORPORALS | | | | | |
| 13 | 28.03.53 | SK14705 | PREISWERK | Gustave Jack | 25 Field Detention Barracks (also listed as PREISWERCK) |
| 7 | 01.06.53 | SG9634 | STOCKFORD | Lawson C. | 1st Comwel Division Provost Company |
| PRIVATE | | | | | |
| 40 | 03.10.53 | SA800036 | HANSON | James Vincent | HQ 1st Commonwealth Division (also listed as 1st Canadian Z Postal Unit) |

| Canadian/ London Gazette issue No. | Canadian/ London Gazette date | Personal Service number | Name | Christian names | Remarks |
|---|---|---|---|---|---|
| ROYAL CANADIAN ARMY DENTAL CORPS | | | | | |
| MENTION IN DESPATCHES (FOR DISTINGUISHED SERVICES) BRONZE OAK LEAF EMBLEM | | | | | |
| MAJOR | | | | | |
| 40 | 03.10.53 | ZB3185 | FERGUSON | Howard Allan | 25 Field Dental Unit |
| CAPTAINS | | | | | |
| 7 | 13.02.54 | ZB3461 | LAVOIE | Francois Benoit | 25 Field Dental Unit |
| 7 | 13.02.54 | ZK3514 | McPHERSON | Angus Alexander | 59 Field Dental Unit (also listed as 59 Field Sqdn. RCE) |

| Canadian/ London Gazette issue No. | Canadian/ London Gazette date | Personal Service number | Name | Christian names | Remarks |
|---|---|---|---|---|---|
| CANADIAN ARMY INTELLIGENCE CORPS | | | | | |
| MENTION IN DESPATCHES (FOR GALLANT AND DISTINGUISHED SERVICES) | | | | | |
| BRONZE OAK LEAF EMBLEM | | | | | |
| MAJOR | | | | | |
| 7 | 01.06.53 | ZG2243 | McSHANE | King George | No.1 Canadian Field Security, HQ 1st Commonwealth Division |
| STAFF SERGEANT | | | | | |
| 40 | 06.10.51 | SB15152 | HENDERSON | Robert | ← Unit not listed |

| Canadian/ London Gazette issue No. | Canadian/ London Gazette date | Personal Service number | Name | Christian names | Remarks |
|---|---|---|---|---|---|
| CANADIAN POSTAL CORPS | | | | | |
| MENTION IN DESPATCHES (FOR DISTINGUISHED SERVICES) BRONZE OAK LEAF EMBLEM | | | | | |
| PRIVATES | | | | | |
| 7 | 01.06.53 | SD802318 | HAMPEL | Vernon Robert | 25 Canadian Postal Detachment |
| 40 | 03.10.53 | SA800036 | HANSON | James Vincent | 1st Canadian Z Postal Unit (also listed as Canadian Military Provost Corps, 1st Commonwealth Div.) |

ROYAL CANADIAN AIR FORCE

| Canadian/ London Gazette issue No. | Canadian/ London Gazette date | Personal Service number | Name | Initials | Remarks |
|---|---|---|---|---|---|
| ROYAL CANADIAN AIR FORCE | | | | | |
| OBE: OFFICER OF THE ORDER OF THE BRITISH EMPIRE | | | | | |
| WING COMMANDER | | | | | |
| | | 19523 | MUSSELLS | C.H. | ∢ Unit not listed |
| MBE: MEMBER OF THE ORDER OF THE BRITISH EMPIRE | | | | | |
| SQUADRON LEADER | | | | | |
| | | 1874 | LORD | W.H. | ∢ |
| DFC: DISTINGUISHED FLYING CROSS | | | | | |
| FLIGHT LIEUTENANT | | | | | |
| | | 17484 | GLOVER | E.A. | ∢ |
| AFC: AIR FORCE CROSS | | | | | |
| WING COMMANDER | | | | | |
| | | 19984 | MORRISON | H.A. | ∢ |
| SQUADRON LEADER | | | | | |
| | | 19812 | DICKSON | J.D. | ∢ |
| FLIGHT LIEUTENANTS | | | | | |
| | | 30018 | EDWARDS | R.M. | ∢ |
| | | 26365 | PAYNE | D.M. | ∢ |
| AFM: AIR FORCE MEDAL | | | | | |
| CORPORAL | | | | | |
| | | 25233 | REED | G.R. | ∢ |
| BEM: BRITISH EMPIRE MEDAL | | | | | |
| FLIGHT SERGEANT | | | | | |
| | | 2412 | ENGELBERT | A.L. | ∢ |
| CORPORAL | | | | | |
| | | 24068 | TRUDEL | J.B.P.A. | ∢ |
| QUEEN'S COMMENDATION FOR VALUABLE SERVICES IN THE AIR (*) | | | | | |
| BRONZE OAK LEAF EMBLEM | | | | | |
| WING COMMANDER | | | | | |
| | | 21047 | McNAIR | R.W. | ∢ |

∢ Unit not listed

| Canadian/ London Gazette issue No. | Canadian/ London Gazette date | Personal Service number | Name | Initials | Remarks |
|---|---|---|---|---|---|
| QUEEN'S COMMENDATION FOR VALUABLE SERVICES IN THE AIR (*) (continued) | | | | | |
| BRONZE OAK LEAF EMBLEM | | | | | |
| FLIGHT LIEUTENANTS | | | | | |
| | | 23526 | ENDERSBE | C.E. | ∈ Unit not listed |
| | | 6552 | FINKELSTEIN | A. | ∈ |
| | | 17600 | MILLER | J.B. | ∈ |
| | | 17706 | RATCLIFFE | R.E.D. | ∈ |
| | | 20405 | SMITH | W. | ∈ |
| | | 25486 | WOLKOWSKI | E.R. | ∈ |
| FLYING OFFICER | | | | | |
| | | 25597 | WILSON | J.P. | ∈ |
| SERGEANTS | | | | | |
| | | 2554 | BOWMAN | F.S.M. | ∈ |
| | | 25880 | HOWARD | G. | ∈ |
| | | 23347 | POTEKAL | L.C. | ∈ |
| | | 22223 | McKNIGHT | W.S. | ∈ |

(*) The official Canadian listings do not make it clear whether of not Queen's Commendations were awarded for operational or non-operational flying duties.

## AWARDS CONFERRED BY THE PRESIDENT OF THE UNITED STATES OF AMERICA

| Canadian/ London Gazette issue No. | Canadian/ London Gazette date | Personal Service number | Name | Initials | Remarks |
|---|---|---|---|---|---|
| DFC: DISTINGUISHED FLYING CROSS | | | | | |
| WING COMMANDER | | | | | |
| 234/53 | | 20361 | LINDSAY, DFC | J.D. | ∈ |
| GROUP CAPTAIN | | | | | |
| 55/53 | | 19514 | HALE, DFC, CD | E.B. | ∈ |
| FLIGHT LIEUTENANTS | | | | | |
| 55/53 | | 120772 | FLEMING | S.B. | ∈ |
| 153/53 | | 17484 | GLOVER, DFC | E.A. | ∈ |
| 153/53 | | 30003 | LAFRANCE | J.C.A. | ∈ |
| BRONZE STAR | | | | | |
| SQUADRON LEADER | | | | | |
| 490/51 | | 16928 | REED, DFC | J.T. | ∈ |
| AIR MEDAL | | | | | |
| WING COMMANDER | | | | | |
| 742/53 | | 20465 | DAVIDSON, DFC, CD | R.T.P. | ∈<br>∈ Unit not listed |

| Canadian/ London Gazette issue No. | Canadian/ London Gazette date | Personal Service number | Name | Initials | Remarks |
|---|---|---|---|---|---|

AWARDS CONFERRED BY THE PRESIDENT OF THE UNITED STATES OF AMERICA

AIR MEDAL (continued)

| Canadian/ London Gazette issue No. | Canadian/ London Gazette date | Personal Service number | Name | Initials | Remarks |
|---|---|---|---|---|---|
| SQUADRON LEADERS | | | | | |
| 362/54 | | 19703 | FOX | W.W. | ← Unit not listed |
| 362/54 | | 19727 | MacKAY, DFC | J. | ← |
| FLIGHT LIEUTENANTS | | | | | |
| 362/54 | | 17822 | BLISS | W.H.F. | ← |
| 362/54 | | 33697 | CAREW | R.D. | ← |
| 742/53 | | 10062 | EVANS, DFC, CD | F.W. | ← |
| 284/53 | | 17923 | LOWRY | R.E. | ← |
| 742/53 | | 30004 | NICHOLS | G.H. | ← |
| FLYING OFFICERS | | | | | |
| 742/53 | | 49644 | LAMBROS, DFC | A. | ← |
| 55/53 | | 14565 | NIXON | W.G. | ← |

← Unit not listed

SUPPORTING SERVICES TO THE CANADIAN ARMED FORCES - KOREA

| Canadian/ London Gazette issue No. | Canadian London Gazette date | Personal Service number | Name | Initials | Remarks |
|---|---|---|---|---|---|
| YMCA | | | | | |
| MBE: MEMBER OF THE ORDER OF THE BRITISH EMPIRE | | | | | |
| not listed | | | MISS STRONACH | N.E. | |

| Canadian/ London Gazette issue No. | Canadian/ London Gazette date | Personal Service number | Name | Initials | Remarks |
|---|---|---|---|---|---|

SALVATION ARMY

MBE: MEMBER OF THE ORDER OF THE BRITISH EMPIRE

DEPUTY ASSISTANT COMMISSIONER

| Canadian/ London Gazette issue No. | Canadian/ London Gazette date | Personal Service number | Name | Initials | Remarks |
|---|---|---|---|---|---|
| not listed | | B4/473 | SEMMENS | J.C. | |

AUSTRALIA

| London Gazette issue No. | London Gazette date | Personal Service number | Name | Christian names | Remarks |
|---|---|---|---|---|---|
| ROYAL AUSTRALIAN NAVY | | | | | |
| CBE: COMMANDER OF THE ORDER OF THE BRITISH EMPIRE | | | | | |
| COMMODORE | | | | | |
| 39682 | 28.10.52 | 0493 | HARRIES, ADC | David Hugh | HMAS Sydney |
| DSO: DISTINGUISHED SERVICE ORDER | | | | | |
| CAPTAINS | | | | | |
| 39336 | 18.09.51 | 058 | BECHER, DSC | Otto Humphrey | HMAS Warramunga |
| 39870 | 01.06.53 | 0407 | GATACRE, DSC | Galfrey George Ormonde | HMAS Anzac |
| OBE: OFFICER OF THE ORDER OF THE BRITISH EMPIRE | | | | | |
| LIEUTENANT COMMANDER (E) | | | | | |
| | | not listed | BULLEN | D.O. | not listed |
| 2ND BAR TO DSC: DISTINGUISHED SERVICE CROSS | | | | | |
| COMMANDER | | | | | |
| 39275 | 23.12.52 | 0105 | BRACEGIRDLE, DSC | Warwick Seymour | HMAS Bataan |
| BAR TO DSC: DISTINGUISHED SERVICE CROSS | | | | | |
| LIEUTENANT COMMANDER | | | | | |
| 39578 | 20.06.52 | 0425 | GLADSTONE, DSC | Geoffrey Vernon | HMAS Warramunga |
| DSC: DISTINGUISHED SERVICE CROSS | | | | | |
| COMMANDERS | | | | | |
| 39578 | 20.06.52 | 0723 | MARKS | William Beresford Moffat | HMAS Bataan |
| 39725 | 23.12.52 | 0952 | RAMSAY | James Maxwell | HMAS Warramunga |
| LIEUTENANT COMMANDERS | | | | | |
| 39682 | 28.10.52 | E1330 | BOWLES | Walter George | HMAS Sydney |
| 40130 | 23.03.54 | 0204 | CLARKE | Domara Andrews Heap | HMAS Culgoa |
| 39687 | 28.10.52 | 0302 | DOLLARD | Allen Nelson | HMAS Murchison |
| 40011 | 06.11.53 | 0986 | ROBERTS | William Owen Chellew | HMAS Anzac (also listed: HMAS Murchison) |
| 39885 | 12.06.53 | 01030 | SAVAGE | Robert Cecil | HMAS Condamine |

| London Gazette issue No. | London Gazette date | Personal Service number | Name | Christian names | Remarks |
|---|---|---|---|---|---|
| DSC: DISTINGUISHED SERVICE CROSS (continued) | | | | | |
| LIEUTENANTS | | | | | |
| 39682 | 28.10.52 | 027 | BAILEY | Harold Edwin | HMAS Sydney |
| 39682 | 28.10.52 | 0654 | BEANGE | Guy Alexander | HMAS Sydney |
| 39682 | 28.10.52 | 0625 | KELLY | James Maxwell | HMAS Murchison |
| | | 0987 | ROBERTSON | A.J. | HMAS Anzac |
| MBE: MEMBER OF THE ORDER OF THE BRITISH EMPIRE | | | | | |
| COMMISSIONED ENGINEER | | | | | |
| 39870 | 01.06.53 | 0182 | CARTER | Jack Broadley | HMAS Condamine |
| DSM: DISTINGUISHED SERVICE MEDAL | | | | | |
| CHIEF PETTY OFFICER | | | | | |
| 39725 | 19.12.52 | F24152 | HARRIS | Alfred Sydney | HMAS Warramunga |
| 39326 | 25.05.51 | F22772 | ROE | William Arthur | HMAS Bataan |
| OBSERVER II | | | | | |
| 39682 | 28.10.52 | A40205 | HUGHES | Gordon Churchill | HMAS Sydney |
| LEADING SEAMAN | | | | | |
| 39578 | 20.06.52 | S35344 | ADAMS | Alan Trevor | HMAS Sydney (also listed: HMAS Warramunga) |
| BEM: BRITISH EMPIRE MEDAL | | | | | |
| CHIEF ENGINE ROOM ARTIFICER | | | | | |
| 39336 | 18.09.51 | A23509 | BOYD | Jack | HMAS Warramunga |
| CHIEF ORDNANCE ARTIFICER | | | | | |
| 40011 | 06.11.53 | 29576 | DALGLEISH | Andrew Cloag | HMAS Anzac |
| ELECTRICAL ARTIFICER (3) | | | | | |
| 39885 | 12.06.53 | 40064 | BAXTER | James | HMAS Anzac |
| CHIEF PETTY OFFICER | | | | | |
| 39725 | 23.12.52 | 24157 | HARRIS | Alfred Sydney | HMAS Warramunga |
| MENTION IN DESPATCHES (FOR GALLANT AND DISTINGUISHED SERVICES) | | | | | |
| BRONZE OAK LEAF EMBLEM | | | | | |
| COMMANDER | | | | | |
| 39326 | 25.05.51 | 0723 | MARKS | William Beresford Moffat | HMAS Bataan |
| LIEUTENANT-COMMANDER | | | | | |
| 39682 | 28.10.52 | not listed | MURRAY | Brian Stewart | HMAS Sydney (also listed: initials T.W.) |

| London Gazette issue No. | London Gazette date | Personal Service number | Name | Christian names | Remarks |
|---|---|---|---|---|---|
| MENTION IN DESPATCHES (FOR GALLANT AND DISTINGUISHED SERVICES) (continued) | | | | | |
| BRONZE OAK LEAF EMBLEM | | | | | |
| LIEUTENANTS | | | | | |
| 39578 | 20.06.52 | not listed | BARNETT | Kenneth Malcolm | HMAS Warramunga |
| 39682 | 28.10.52 | not listed | BROWN, DFC | George Firth Spencer | HMAS Sydney |
| 39336 | 18.09.51 | 0147 | BURNETT | Patrick Richard | HMAS Bataan |
| 40011 | 06.11.53 | not listed | DOYLE | Peter Hogath | HMAS Anzac (also listed: HMAS Sydney) |
| | | not listed | GOLDER | John Williams | HMAS Bataan |
| 39682 | 28.10.52 | not listed | GORDON, DFC | Alexander Hughie | HMAS Sydney |
| 39725 | 23.12.52 | not listed | HARRINGTON | John Edward | HMAS Warramunga |
| 39682 | 28.10.52 | not listed | SEED | Peter William | HMAS Sydney |
| LIEUTENANT (E) | | | | | |
| 39682 | 28.10.52 | not listed | ROURKE | William John | HMAS Sydney |
| LIEUTENANT (S) | | | | | |
| 39336 | 18.09.51 | not listed | KEMP | William Alexander | HMAS Sydney (also listed: HMAS Bataan) |
| SENIOR COMMISSIONED ELECTRICAL OFFICER (L) | | | | | |
| 39336 | 18.09.51 | | MR. CANTOR | Samuel George | HMAS Bataan |
| SUB-LIEUTENANT | | | | | |
| 39682 | 28.10.52 | not listed | ROLAND | Armand John | HMAS Sydney |
| CHIEF ENGINE ROOM ARTIFICERS | | | | | |
| 40011 | 06.11.53 | 28813 | GRUBNAU | Roy | HMAS Anzac |
| 40011 | 06.11.53 | 34618 | PACEY | Frederick James | HMAS Condamine |
| CHIEF PETTY OFFICER STOKER MECHANIC | | | | | |
| 39578 | 20.06.52 | 23447 | DIX | Charles Edward | HMAS Melbourne (also listed: HMAS Warramunga) |
| 39336 | 18.09.51 | 22923 | FORBES | William Chalmers | HMAS Bataan |
| CHIEF ELECTRICAL ARTIFICER | | | | | |
| 39682 | 28.10.52 | 25984 | WHELAN | James Patrick | HMAS Sydney |
| CHIEF AIRMAN | | | | | |
| 39682 | 28.10.52 | A40340 | GARDNER | William Daniel | HMAS Sydney |
| CHIEF PETTY OFFICERS | | | | | |
| 39578 | 20.06.52 | 22174 | BACKEN | John William | HMAS Sydney (also listed: HMAS Warramunga) |
| | | 38549 | ERRINGTON | Robert Coad | HMAS Melbourne (also listed: HMAS Bataan) |
| 39682 | 28.10.52 | 23151 | FERNANDEZ | Eugene Elderfield | HMAS Sydney (also listed: FARNANDEZ) |
| 39336 | 18.09.51 | 20830 | HAYNES | Cyril | HMAS Bataan |

| London Gazette issue No. | London Gazette date | Personal Service number | Name | Christian names | Remarks |
|---|---|---|---|---|---|
| MENTION IN DESPATCHES (FOR GALLANT AND DISTINGUISHED SERVICES) (continued) | | | | | |
| BRONZE OAK LEAF EMBLEM | | | | | |
| CHIEF PETTY OFFICER TELEGRAPHIST | | | | | |
| 39725 | 23.12.52 | 29898 | STRINGER | Cyril Robert | HMAS Warramunga |
| PETTY OFFICER COOK (S) | | | | | |
| 39336 | 18.09.51 | 21518 | O'CONNOR | John | HMAS Sydney (also listed: HMAS Bataan) |
| LEADING STOKER MECHANIC | | | | | |
| 39725 | 23.12.52 | 35303 | McNAMARA | Robert George | HMAS Bataan |
| LEADING SEAMAN | | | | | |
| 39725 | 23.12.52 | 28958 | SMITH | Stanley Rowland | HMAS Bataan |
| YEOMEN OF SIGNALS | | | | | |
| 39870 | 01.06.53 | 29231 | JONES | William Alfred | HMAS Condamine |
| 40130 | 23.03.54 | 27602 | MACRAE | Norman Bruce Dickson | HMAS Tobruk |
| ENGINE ROOM ARTIFICER (3) | | | | | |
| 40130 | 23.03.54 | 38211 | MARKHAM | John Kevin | HMAS Culgoa |
| ELECTRICIAN (E) | | | | | |
| | | 40166 | CANNON | Herbert George | HMAS Bataan |

## AWARDS CONFERRED BY THE PRESIDENT OF THE UNITED STATES OF AMERICA

| London Gazette issue No. | London Gazette date | Personal Service number | Name | Christian names | Remarks |
|---|---|---|---|---|---|
| LEGION OF MERIT, DEGREE OF OFFICER | | | | | |
| REAR-ADMIRAL | | | | | |
| 40406 | 11.02.55 | 0493 | HARRIES, CBE | David Hugh | HMAS Sydney |
| CAPTAINS | | | | | |
| 40406 | 11.02.55 | 058 | BECHER, DSO, DSC | Otto Humphrey | HMAS Warramunga |
| 40406 | 11.02.55 | 0914 | PEEK, OBE, DSC | Richard Innes | HMAS Tobruk |
| COMMANDERS | | | | | |
| | | 0105 | BRACEGIRDLE, DSC | Warwick Seymour | HMAS Bataan |
| 40406 | 11.02.55 | 0302 | DOLLARD, DSC | Allen Nelson | HMAS Murchison |
| | | 0952 | RAMSAY | James Maxwell | HMAS Warramunga |
| LIEUTENANT-COMMANDER | | | | | |
| 40406 | 11.02.55 | E1330 | BOWLES, DSC | Walter George | HMAS Sydney |
| LEGION OF MERIT, DEGREE OF LEGIONNAIRE | | | | | |
| COMMANDER | | | | | |
| | | 0723 | MARKS | William Beresford Moffat | HMAS Bataan |
| BRONZE STAR | | | | | |
| LIEUTENANT-COMMANDER | | | | | |
| 40406 | 11.02.55 | 0425 | GLADSTONE, DSC | Geoffrey Vernon | HMAS Warramunga |

AUSTRALIAN MILITARY

| London Gazette issue No. | London Gazette date | Personal Service number | Name | Christian names | Remarks |
|---|---|---|---|---|---|
| HEADQUARTERS STAFF | | | | | |
| CBE: COMMANDER OF THE ORDER OF THE BRITISH EMPIRE | | | | | |
| BRIGADIERS | | | | | |
| 39862 | 26.05.53 | 4/5 | DALY, DSO, OBE | Thomas Joseph | Staff Corps: late Infantry; Commander 28th Comwel Inf Brigade |
| | 01.06.54 | 2/16 | WILTON | John Gordon N. | Staff Corps: late Infantry; Commander 28th Comwel Inf Brigade |
| OBE: OFFICER OF THE ORDER OF THE BRITISH EMPIRE | | | | | |
| MAJORS | | | | | |
| 40025 | 24.11.53 | 2/260 | NEWTON | Colin Elwell McFarquhar | Staff Corps |
| 39862 | 26.05.53 | 2/180 | WHITE | John Francis | 28th Comwel Inf Brigade |
| MBE: MEMBER OF THE ORDER OF THE BRITISH EMPIRE | | | | | |
| MAJOR | | | | | |
| | | 31324 | STRETTON | A.B. | 28th Comwel Inf Brigade |
| CAPTAINS | | | | | |
| 40025 | 24.11.53 | 2/508 | MORRISON | Alan Lindsay | 28th Comwel Inf Brigade |
| 40025 | 24.11.53 | 1/120 | WEIR | Donald David | Staff Corps |
| LIEUTENANT QUARTERMASTER | | | | | |
| 40025 | 24.11.53 | 3/267 | HOSKING | Hubert Arthur James | Royal Australian Infantry Corps |
| MC: MILITARY CROSS | | | | | |
| LIEUTENANT | | | | | |
| 39862 | 26.05.53 | 5/7012 | SKIPPER | Jack Harold | 28th Comwel Inf Brigade |
| MENTION IN DESPATCHES (FOR GALLANT AND DISTINGUISHED FLYING SERVICES - POSTHUMOUS) | | | | | |
| BRONZE OAK LEAF EMBLEM | | | | | |
| CAPTAIN | | | | | |
| 39661 | 03.10.52 | 5/7003 | LUSCOMBE | B.T. | Royal Australian Artillery |
| MENTION IN DESPATCHES (FOR GALLANT AND DISTINGUISHED FLYING SERVICES) | | | | | |
| BRONZE OAK LEAF EMBLEM | | | | | |
| CAPTAIN | | | | | |
| 40012 | 10.11.53 | 1/7007 | DECON | R.S. | Royal Australian Artillery |

| London Gazette issue No. | London Gazette date | Personal Service number | Name | Christian names | Remarks |
|---|---|---|---|---|---|
| MENTION IN DESPATCHES (FOR GALLANT AND DISTINGUISHED SERVICES) | | | | | |
| BRONZE OAK LEAF EMBLEM | | | | | |
| SERGEANT | | | | | |
| 40025 | 24.11.53 | 5/1357 | PICK | J.I. | Royal Australian Infantry Corps BCFK |
| CORPORAL | | | | | |
| | | 3/2239 | LANG | G.A. | HQ 1st Commonwealth Division |

AWARDS CONFERRED BY THE PRESIDENT OF THE UNITED STATES OF AMERICA

| London Gazette issue No. | London Gazette date | Personal Service number | Name | Christian names | Remarks |
|---|---|---|---|---|---|
| LEGION OF MERIT, DEGREE OF CHIEF COMMANDER | | | | | |
| LIEUTENANT-GENERAL | | | | | |
| 39412 | 18.12.51 | 3/4 | ROBERTSON, KBE, DSO | Sir Henry Clement Hugh | Staff Corps: BCOF |
| LEGION OF MERIT, DEGREE OF COMMANDER | | | | | |
| LIEUTENANT-GENERAL | | | | | |
| | | 3/6 | BRIDGEFORD | W. | Staff Corps: BCFK |
| 40325 | 12.11.54 | 3/27 | WELLS, CB, DSO, OBE | Henry | Staff Corps: BCFK |
| LEGION OF MERIT, DEGREE OF OFFICER | | | | | |
| BRIGADIERS | | | | | |
| 39999 | 27.10.53 | 4/5 | DALY, CBE, DSO | Thomas Joseph | Staff Corps: late Infantry; Commander 28th Comwel Inf Brigade |
| | | 2/16 | WILTON | John Gordon N. | Staff Corps: late Infantry; Commander 28th Comwel Inf Brigade |
| LEGION OF MERIT, DEGREE OF LEGIONNAIRE | | | | | |
| LIEUTENANT-COLONELS | | | | | |
| 40406 | 11.02.55 | 3/37502 | KING, OBE, ED | Leslie Dudley | Staff Corps: BCFK |
| 39456 | 05.02.52 | 1/8001 | MARSON, DSO, ED | Richard Harold | British Commonwealth Liaison Group |
| BRONZE STAR | | | | | |
| CAPTAIN | | | | | |
| 39999 | 27.10.53 | 4/8024 | TABE | Frank Leslie | British Commonwealth Liaison Group |

| London Gazette issue No. | London Gazette date | Personal Service number | Name | Christian names | Remarks |
|---|---|---|---|---|---|

AWARD CONFERRED BY THE PRESIDENT OF THE REPUBLIC OF KOREA

ORDER OF MILITARY MERIT TAIGUK

LIEUTENANT-GENERAL

| London Gazette issue No. | London Gazette date | Personal Service number | Name | Christian names | Remarks |
|---|---|---|---|---|---|
| 39456 | 05.02.52 | 3/4 | ROBERTSON, KBE, DSO | Sir Henry Clement Hugh | Staff Corps: BCOF |

| London Gazette issue No. | London Gazette date | Personal Service number | Name | Christian names | Remarks |
|---|---|---|---|---|---|

ROYAL AUSTRALIAN ARTILLERY

MENTION IN DESPATCHES (FOR GALLANT AND DISTINGUISHED SERVICES)
BRONZE OAK LEAF EMBLEM

MAJOR

| | | | | | |
|---|---|---|---|---|---|
| 39862 | 26.05.53 | 2/109 | WATT | A.D. | attached Royal New Zealand Field Regiment |

LANCE-BOMBARDIER

| | | | | | |
|---|---|---|---|---|---|
| 40012 | 10.11.53 | 203629 | CONWAY | K.L. | ← unit not listed |

GUNNER

| | | | | | |
|---|---|---|---|---|---|
| 40012 | 10.11.53 | 204277 | TAIMAN | H.P. | ← unit not listed |

| London Gazette issue No. | London Gazette date | Personal Service number | Name | Christian names | Remarks |
|---|---|---|---|---|---|

ROYAL AUSTRALIAN CORPS OF ENGINEERS

MC: MILITARY CROSS

CAPTAIN

| | | | | | |
|---|---|---|---|---|---|
| 40025 | 24.11.53 | 2/502 | HUTCHESON | John Malcolm | 28th Field Engineer Regiment |

MENTION IN DESPATCHES (FOR GALLANT AND DISTINGUISHED SERVICES)

BRONZE OAK LEAF EMBLEM

CAPTAIN

| | | | | | |
|---|---|---|---|---|---|
| 39862 | 26.05.53 | 2/494 | GILMORE | I.G.C. | 28th Field Engineer Regiment |

AWARD CONFERRED BY THE PRESIDENT OF THE UNITED STATES OF AMERICA

BRONZE STAR

CAPTAIN

| | | | | | |
|---|---|---|---|---|---|
| 30505 | 01.04.52 | WX35410 | SMITH | Neville Dixon | 28th Field Engineer Regiment |

| London Gazette issue No. | London Gazette date | Personal Service number | Name | Christian names | Remarks |
|---|---|---|---|---|---|
| ROYAL AUSTRALIAN REGIMENT | | | | | |
| GC: GEORGE CROSS | | | | | |
| PRIVATE | | | | | |
| | | 2/400186 | MADDEN | H.W. | 3rd Bn |
| DSO: DISTINGUISHED SERVICE ORDER | | | | | |
| LIEUTENANT-COLONELS | | | | | |
| 39891 | 19.06.53 | 1/10 | AUSTIN | Maurice | 1st Bn Officer Commanding |
| 39233 | 22.05.51 | 2/37504 | FERGUSON, MC | Ian Bruce | 3rd Bn Officer Commanding (service number also listed as 507) |
| 39453 | 01.02.52 | 2/27 | HASSETT, OBE | Francis George | 3rd Bn Officer Commanding |
| 39862 | 26.05.53 | 7/2 | HUGHES | Ronald Lawrence | 3rd Bn Officer Commanding |
| MAJORS | | | | | |
| 39545 | 20.05.52 | 5/400076 | GERKE | Jack | 3rd Bn |
| 39796 | 10.03.53 | 2/285 | MANN | Adrian Smith | 1st Bn |
| OBE: OFFICER OF THE ORDER OF THE BRITISH EMPIRE | | | | | |
| LIEUTENANT-COLONELS | | | | | |
| 39862 | 26.05.53 | 2/37602 | HUTCHISON, DSO,MC,ED | Ian | 1st Bn Officer Commanding (also listed as HUTCHINSON) |
| 40025 | 14.11.53 | 2/47 | LARKIN | George Frederick | 2nd Bn Officer Commanding |
| MAJORS | | | | | |
| 39670 | 14.10.52 | 2/134 | CAREY | John William | 3rd Bn |
| 39862 | 26.05.53 | 5/16 | FINLAYSON | William John | 3rd Bn |
| 40205 | 15.06.54 | 3/210 | HENDERSON | William George | 3rd Bn |
| 39862 | 26.05.53 | 3/314 | MORROW | William John | 3rd Bn |
| 40025 | 24.11.53 | 2/274 | NORRIE | James William | 3rd Bn |
| 39862 | 26.05.53 | 2/258 | SHARP | Derek Granville | 1st Bn |
| 40025 | 24.11.53 | 6/11 | THOMAS, MC | Kevan Brittan | 2nd Bn |
| 40025 | 24.11.53 | 2/37521 | WILSON | Thomas Harry | 2nd Bn |
| 40025 | 24.11.53 | 3/37523 | WOODHOUSE | John Frederick Rance | 2nd Bn |
| MBE: MEMBER OF THE ORDER OF THE BRITISH EMPIRE | | | | | |
| CAPTAINS | | | | | |
| 39265 | 22.06.51 | 3/37571 | CALLANDER | Jack Warwick | 3rd Bn |
| 40205 | 15.06.54 | 3/37686 | HARRINGTON | William Abraham | 3rd Bn |
| 40025 | 24.11.53 | 2/37608 | WELLS | John Bryden | 2nd Bn |
| 39703 | 25.12.52 | 1/8024 | WHALLEY | John Reginald | 3rd Bn |

| London Gazette issue No. | London Gazette date | Personal Service number | Name | Christian names | Remarks |
|---|---|---|---|---|---|
| MBE: MEMBER OF THE ORDER OF THE BRITISH EMPIRE (continued) | | | | | |
| LIEUTENANT | | | | | |
| 40205 | 24.11.53 | 3/1068 | CONNELL | John William Martin | 2nd Bn |
| LIEUTENANT QUARTERMASTER | | | | | |
| 39891 | 19.06.53 | 2/341 | McDERMOTT | Keith Victor Patrick | 1st Bn |
| WARRANT OFFICERS CLASS II | | | | | |
| 39862 | 26.05.53 | 2/1061 | BRENNAN | Leslie Edward | 1st Bn |
| | | 1/205 | HUMPHRIS | A. | 2nd Bn |
| | | 3/884 | MOORE | L.A. | 2nd Bn |
| | | 4/220 | NITSCHKE | E.S. | 1st Bn |
| BAR TO MC: MILITARY CROSS | | | | | |
| CAPTAIN | | | | | |
| 39703 | 25.11.52 | 1/8026 | NICHOLLS, MC | Henry William | 3rd Bn |
| MC: MILITARY CROSS | | | | | |
| MAJORS | | | | | |
| 39891 | 19.06.53 | 3/391 | HEARN | Bruce Barrington | 1st Bn |
| 39891 | 19.06.53 | 5/25 | KAYLER-THOMPSON | Clarence David | 1st Bn |
| 39703 | 25.11.52 | 3/395 | SHELTON | Jeffrey James | 3rd Bn |
| 39661 | 03.10.52 | 3/328 | THOMPSON | David Scott | 1st Bn |
| CAPTAINS | | | | | |
| 39597 | 15.07.52 | 1/116 | CLARK | Lawrence George | 3rd Bn |
| 39205 | 17.04.51 | Z/400335 | DENNESS | Archer Paterson | 3rd Bn |
| 39703 | 25.11.52 | 1/400148 | DODDRELL | Arthur Sydney Roy | 3rd Bn |
| 39670 | 14.10.52 | 2/400352 | KEYS | Alexander George | 3rd Bn (initials also shown as A.G.W.) |
| 39862 | 26.05.53 | 5/57 | RICHARDSON | Rupert Peter | 3rd Bn |
| 39862 | 26.05.53 | 2/37619 | WATERTON | John Thomas Max Charles | 3rd Bn |
| LIEUTENANTS | | | | | |
| 39891 | 19.06.53 | 3/35021 | BOUSFIELD | Brian Nicholson | 3rd Bn |
| 40205 | 15.06.54 | 4/7538 | FORBES | Patrick Oliver Giles | 2nd Bn |
| 39597 | 15.07.52 | 4/7001 | HUGHES | James Curnow | 3rd Bn |
| 39791 | 03.03.53 | 3/35036 | JAMES | William Brian | 1st Bn |
| 40025 | 24.11.53 | 5/7015 | LLOYD | David Ferrers | 3rd Bn |
| 39862 | 26.05.53 | 3/40105 | LUCAS | Gilmer John | 1st Bn |
| 39312 | 17.08.51 | 4/40059 | MONTGOMERIE | Leonard | 3rd Bn |
| 39703 | 25.11.52 | 2/35017 | PEARS | Maurice Bertram | 3rd Bn |
| 39597 | 15.07.52 | 1/7003 | PEMBROKE | Arthur Thomas | 3rd Bn |

| London Gazette issue No. | London Gazette date | Personal Service number | Name | Christian names | Remarks |
|---|---|---|---|---|---|
| **MC: MILITARY CROSS (continued)** | | | | | |
| **LIEUTENANTS (continued)** | | | | | |
| 39703 | 25.11.52 | 2/35018 | STEWART | James David | 3rd Bn |
| 39891 | 19.06.53 | 3/10128 | WILLIAMS | Dennis North | 1st Bn |
| 40025 | 24.11.53 | 5/7013 | YACOPETTI | Charles Peter | 3rd Bn |
| 39484 | 04.03.52 | 2/40101 | YOUNG | James Hay-Archer | 3rd Bn |
| | | | | | |
| **BAR TO DCM: DISTINGUISHED CONDUCT MEDAL** | | | | | |
| **SERGEANT** | | | | | |
| 39484 | 04.03.52 | 2/400239 | ROWLINSON, DCM | William Josiah | 3rd Bn |
| | | | | | |
| **DCM: DISTINGUISHED CONDUCT MEDAL** | | | | | |
| **WARRANT OFFICER CLASS II** | | | | | |
| 39670 | 14.10.52 | 4/400006 | OPIE | Leonard Murray | 3rd Bn |
| **SERGEANT** | | | | | |
| 39824 | 14.04.53 | 3/1967 | MORRISON | Edward John | 3rd Bn |
| **CORPORAL** | | | | | |
| 39448 | 25.01.52 | 2/400239 | ROWLINSON | William Josiah | 3rd Bn |
| **LANCE-CORPORAL** | | | | | |
| 39597 | 15.07.52 | 1/4000092 | BURNETT | James | 3rd Bn |
| | | | | | |
| **GM: GEORGE MEDAL** | | | | | |
| **SERGEANT** | | | | | |
| 39265 | 19.02.51 | NX135280 | MURRAY | Thomas Mervyn | 3rd Bn |
| | | | | | |
| **BAR TO MM: MILITARY MEDAL** | | | | | |
| **PRIVATE** | | | | | |
| 39947 | 25.08.53 | 3/400816 | WHITE, MM (1945) | Albert Michess | 3rd Bn |
| | | | | | |
| **MM: MILITARY MEDAL** | | | | | |
| **WARRANT OFFICER CLASS II** | | | | | |
| 39703 | 25.11.52 | 2/116 | STANLEY | Arthur George | 3rd Bn |
| **STAFF SERGEANT** | | | | | |
| 39703 | 25.11.52 | 2/2088 | PURVIS | Alfred Donald | 3rd Bn |
| **SERGEANTS** | | | | | |
| 39703 | 25.11.52 | 2/6086 | BROWN | Vincent John Edward | 3rd Bn |
| 39947 | 25.05.53 | 5/1534 | BRUCE | William James Joseph | 2nd Bn |
| 39997 | 27.10.53 | 5/2053 | COOPER | Brian Charles | 2nd Bn |
| 39891 | 19.06.53 | 2/401090 | DAVIES | Cecil Ernest | 3rd Bn |

| London Gazette issue No. | London Gazette date | Personal Service number | Name | Christian names | Remarks |
|---|---|---|---|---|---|
| MM: MILITARY MEDAL | | | | | |
| SERGEANTS (continued) | | | | | |
| 40205 | 15.06.54 | 2/401127 | DIGGS | Thomas | 3rd Bn |
| 39670 | 14.10.52 | 2/400234 | EVELEIGH | Cecil James | 3rd Bn |
| 39796 | 10.03.53 | 2/4880 | McNULTY | Edward John | 1st Bn |
| 39670 | 14.10.52 | 1/400133 | O'CONNELL | Patrick John | 3rd Bn |
| 39703 | 25.11.52 | 3/2053 | STRONG | Rex William | 3rd Bn |
| CORPORALS | | | | | |
| 39545 | 20.05.52 | 3/400006 | BLACK | John Kenneth | 3rd Bn |
| 39597 | 15.07.52 | 1/400095 | BOSWORTH | Edward Fowkes | 3rd Bn |
| 39703 | 25.11.52 | 5/1461 | CAMERON | Donald George | 3rd Bn |
| 39968 | 22.09.53 | 2/2913 | CASHMAN | Ronald Kenneth | 3rd Bn |
| 39318 | 17.08.51 | 4/158 | DAVIE | Donald Breynard | 3rd Bn |
| 39862 | 26.05.53 | 3/400353 | JUBB | Thomas John | 3rd Bn |
| 39862 | 26.05.53 | 2/400620 | McCRINDLE | Ronald John | 1st Bn |
| 39824 | 14.04.53 | 3/3841 | MACKAY | Francis Leo | 3rd Bn |
| 40205 | 15.06.54 | 1/2303 | MAGUIRE | Thomas William | 3rd Bn |
| 39862 | 26.05.53 | 1/9942 | MENE | Charlie | 1st Bn |
| 39545 | 20.05.52 | 1/400149 | PARK | John | 3rd Bn |
| 39448 | 25.01.52 | 5/40049 | PARRY | Ray Norman | 3rd Bn |
| 39862 | 26.05.53 | 2/4588 | SAVILLE | Bruce | 3rd Bn |
| 39661 | 03.10.52 | 2/2709 | TAYLOR | Laurence E. | 1st Bn |
| 39862 | 25.05.53 | 2/400495 | THOMAS | Kevin Claude | 3rd Bn |
| 39670 | 14.10.52 | 2/400106 | TUNSTALL | Thomas Gilroy | 3rd Bn |
| LANCE-CORPORALS | | | | | |
| 39997 | 27.10.53 | 1/400604 | CROCKFORD | Kenneth Humber | 2nd Bn |
| 39862 | 26.05.53 | 2/1838 | HOLDEN | Leo Clarence | 3rd Bn |
| 39751 | 13.01.53 | 2/400645 | McCARTHY | David | 1st Bn |
| 39862 | 26.05.53 | 2/400615 | RALSTON | Gavin Carmichael | 1st Bn |
| 40025 | 24.11.53 | 2/10667 | RICHARDSON | Robert | 2nd Bn |
| 39862 | 26.05.53 | 7/73 | WILSON | Maxwell Eric | 3rd Bn |
| PRIVATES | | | | | |
| 39448 | 25.01.52 | 3/2196 | DUNQUE | Ronald Edward | 3rd Bn |
| 40025 | 15.06.54 | 1/400713 | KENT | George Edward | 2nd Bn |
| 40025 | 15.06.54 | 3/2491 | McAULIFFE | James Michael | 2nd Bn |
| 39703 | 25.11.52 | 2/4576 | POWER | David Richard | 3rd Bn |

| London Gazette issue No. | London Gazette date | Personal Service number | Name | Christian names | Remarks |
|---|---|---|---|---|---|
| MM: MILITARY MEDAL | | | | | |
| PRIVATES (continued) | | | | | |
| 39312 | 17.08.51 | 3/400223 | SMITH | Ronald Francis Alfred | 3rd Bn |
| 39682 | 28.10.52 | 1/1745 | WHITE | Alfred | 3rd Bn |
| 39862 | 26.05.53 | 2/400430 | WHITE | Herbert Gordon | 3rd Bn |
| 39891 | 19.06.53 | 6/400104 | WOOLLEY | Henry Arthur | 3rd Bn |
| BEM: BRITISH EMPIRE MEDAL | | | | | |
| WARRANT OFFICER CLASS II | | | | | |
| | | 2/45252 | ANDERSON | I.J. | 3rd Bn |
| | | 1/248 | CALLAHAN | D.E. | not listed |
| SERGEANT | | | | | |
| 39891 | 19.06.53 | 2/10234 | WATKINS | William George | 1st Bn |
| MENTION IN DESPATCHES (FOR GALLANT AND DISTINGUISHED SERVICES - POSTHUMOUS) | | | | | |
| BRONZE OAK LEAF EMBLEM | | | | | |
| LIEUTENANTS | | | | | |
| 39862 | 26.05.53 | 2/37657 | CLIFF | P.H. | 1st Bn |
| 40025 | 24.11.53 | 1/7554 | QUINLAN | J.P. | 3rd Bn (also listed as QUINLAND) |
| 39862 | 26.05.53 | 6/3760 | SEATON | J.L. | 1st Bn |
| 40506 | 07.06.55 | 2/35020 | SMITH | F.C. | 3rd Bn |
| SERGEANT | | | | | |
| 40025 | 24.11.53 | 2/400126 | COCKS | B.K. | 3rd Bn |
| CORPORALS | | | | | |
| 39862 | 26.05.53 | 2/400358 | CARTER | L.A. | 1st Bn |
| 40205 | 15.06.54 | 5/1482 | COOPER | K.J. | 2nd Bn |
| LANCE-CORPORAL | | | | | |
| 40506 | 07.06.55 | 2/401173 | BOURKE | E.G. | 2nd Bn |
| PRIVATE | | | | | |
| 40506 | 07.06.55 | 3/400376 | TERRY | L.J. | 3rd Bn |
| MENTION IN DESPATCHES (FOR GALLANT AND DISTINGUISHED SERVICES WHILST PRISONERS OF WAR IN NORTH KOREA) | | | | | |
| BRONZE OAK LEAF EMBLEM | | | | | |
| LIEUTENANT | | | | | |
| 40334 | 23.11.54 | 5/7013 | YACOPETTI, MC | Charles Peter | 3rd Bn |
| CORPORAL | | | | | |
| 40334 | 23.11.54 | 2/400000 | BUCK | Donald Pattison | not listed |

| London Gazette issue No. | London Gazette date | Personal Service number | Name | Christian names | Remarks |
|---|---|---|---|---|---|
| MENTION IN DESPATCHES (FOR GALLANT AND DISTINGUISHED SERVICES WHILST PRISONERS OF WAR IN NORTH KOREA) | | | | | |
| BRONZE OAK LEAF EMBLEM (continued) | | | | | |
| PRIVATES | | | | | |
| 40334 | 23.11.54 | 3/400024 | GWYTHER | Keith Roy | 3rd Bn |
| 40334 | 23.11.54 | 2/400311 | HOLLIS | Thomas Henry John | 3rd Bn |
| 40334 | 23.11.54 | 2/400030 | PARKER | Robert | 3rd Bn |
| MENTION IN DESPATCHES (FOR GALLANT AND DISTINGUISHED SERVICES) | | | | | |
| BRONZE OAK LEAF EMBLEM | | | | | |
| LIEUTENANT-COLONEL | | | | | |
| 40025 | 24.11.53 | 1/8 | MacDONALD, OBE | A.L. | 3rd Bn Officer Commanding |
| MAJORS | | | | | |
| 39456 | 05.02.52 | 3/40048 | BROWN | W.F. | 3rd Bn |
| 40205 | 15.06.54 | 3/37545 | GRIFF, MC | E.M. | 3rd Bn |
| 39545 | 20.05.52 | 2/257 | HARDIMAN | J. | 3rd Bn |
| 39545 | 20.05.52 | 3/395 | SHELTON | J.J. | 3rd Bn |
| 39891 | 19.06.53 | 2/398 | SMITH | E.H. | 1st Bn |
| 40025 | 24.11.53 | 2/361 | TRENERRY | J.B.M. | 3rd Bn |
| CAPTAINS | | | | | |
| 39456 | 05.02.52 | 5/7004 | BENNETT | P.H. | 3rd Bn |
| 39703 | 25.11.52 | 2/507 | BRUMFIELD | I.R.W. | 3rd Bn |
| 39456 | 05.02.52 | 2/510 | EYLES | L.A. | 3rd Bn |
| 39597 | 15.07.52 | 6/48 | LEARY | G.J. | 3rd Bn |
| 30703 | 25.11.52 | 2/493 | Le MERCIER | R.J.P.J. | 3rd Bn |
| 39862 | 26.05.53 | 3/478 | MOLONEY | J.C.F. | 3rd Bn |
| 40205 | 15.06.54 | 2/40071 | MORAHAN | J.O. | 3rd Bn |
| 39456 | 05.02.52 | 5/7511 | MORAHAN | R.G. | 3rd Bn |
| 39484 | 04.03.52 | 1/8026 | NICHOLLS, MC | H.W. | 3rd Bn |
| 39703 | 25.11.52 | 1/113 | PREECE | A.V. | 3rd Bn |
| 39862 | 26.05.53 | 2/37658 | ROBERTSON | B.E.D. | 3rd Bn |
| 39891 | 19.06.53 | 5/94 | ROGERS | R.B. | 1st Bn |
| LIEUTENANTS | | | | | |
| 39265 | 22.06.51 | 3.35007 | ARGENT | A. | 3rd Bn |
| 39703 | 25.11.52 | 3/40086 | BATTERSBY | R.H. | 3rd Bn (also listed as BETTERSBY) |
| | | 5/7004 | BENNETT | P.H. | 3rd Bn |
| 39862 | 26.05.53 | 2/37656 | BURKE | J.M.D. | 1st Bn |
| 39597 | 15.07.52 | 2/35015 | FALVEY | B.J. | 3rd Bn |

| London Gazette issue No. | London Gazette date | Personal Service number | Name | Christian names | Remarks |
|---|---|---|---|---|---|
| ROYAL AUSTRALIAN ARMY CHAPLAINS' DEPARTMENT | | | | | |
| MBE: MEMBER OF THE ORDER OF THE BRITISH EMPIRE | | | | | |
| CHAPLAINS | | | | | |
| 39703 | 25.11.52 | VX700033 | REVEREND PHILLIPS | Elzear Basil | ← Unit not listed |
| 39862 | 26.05.53 | 1/8037 | REVEREND SHINE | Francis Anthony | attached 1st Bn R.A.R. |

| London Gazette issue No. | London Gazette date | Personal Service number | Name | Christian names | Remarks |
|---|---|---|---|---|---|
| ROYAL AUSTRALIAN ARMY SERVICE CORPS | | | | | |
| BEM: BRITISH EMPIRE MEDAL | | | | | |
| WARRANT OFFICER CLASS II | | | | | |
| 40025 | 24.11.53 | 2/1132 | HARDING | Ernest Frank | ← Unit not listed |
| MENTION IN DESPATCHES (FOR GALLANT AND DISTINGUISHED SERVICES) | | | | | |
| BRONZE OAK LEAF EMBLEM | | | | | |
| SERGEANTS | | | | | |
| 40012 | 10.11.53 | 206738 | COULTER | D. McB. | ← |
| 40012 | 10.11.53 | 206501 | JORDAN | T. | ← |
| PRIVATE | | | | | |
| 39703 | 25.11.52 | QX41226 | MANCKTELOW | D. | attached 3rd Bn R.A.R. |

← Unit not listed

| London Gazette issue No. | London Gazette date | Personal Service number | Name | Christian names | Remarks |
|---|---|---|---|---|---|
| ROYAL AUSTRALIAN ARMY MEDICAL CORPS | | | | | |
| MBE: MEMBER OF THE ORDER OF THE BRITISH EMPIRE | | | | | |
| CAPTAIN | | | | | |
| 39670 | 14.10.52 | SX700053 | BARNES | Robert | attached 3rd Bn R.A.R. |
| MM: MILITARY MEDAL | | | | | |
| CORPORAL | | | | | |
| 39891 | 19.06.53 | 2/4688 | THOMAS | John Francis | attached 1st Bn R.A.R. |
| BEM: BRITISH EMPIRE MEDAL | | | | | |
| SERGEANT | | | | | |
| 40205 | 15.06.54 | 3/749 | MILLER | W. James | attached 3rd Bn R.A.R. |
| MENTION IN DESPATCHES (FOR GALLANT AND DISTINGUISHED SERVICES) | | | | | |
| BRONZE OAK LEAF EMBLEM | | | | | |
| CORPORAL | | | | | |
| 39641 | 09.09.52 | 2/1561 | DOWELL | E.G. | attached 1st Bn R.A.R. |

| London Gazette issue No. | London Gazette date | Personal Service number | Name | Christian names | Remarks |
|---|---|---|---|---|---|

ROYAL AUSTRALIAN ARMY ORDNANCE CORPS

MBE: MEMBER OF THE ORDER OF THE BRITISH EMPIRE

LIEUTENANT

| London Gazette issue No. | London Gazette date | Personal Service number | Name | Christian names | Remarks |
|---|---|---|---|---|---|
| 40025 | 24.11.53 | V325447 | SCHMIDT | Bert Stanley | BCOF |

| London Gazette issue No. | London Gazette date | Personal Service number | Name | Christian names | Remarks |
|---|---|---|---|---|---|
| ROYAL AUSTRALIAN CORPS OF ELECTRICAL AND MECHANICAL ENGINEERS | | | | | |
| BEM: BRITISH EMPIRE MEDAL | | | | | |
| SERGEANT | | | | | |
| 40205 | 15.06.54 | 3/2350 | SUMMERS | James Edward | attached 2nd Bn R.A.R. |
| MENTION IN DESPATCHES (FOR GALLANT AND DISTINGUISHED SERVICES) | | | | | |
| BRONZE OAK LEAF EMBLEM | | | | | |
| SERGEANT | | | | | |
| 40205 | 15.06.54 | 3/2256 | PARRY | J.E. | attached 3rd Bn R.A.R. |

AWARD CONFERRED BY THE PRESIDENT OF THE UNITED STATES OF AMERICA

| London Gazette issue No. | London Gazette date | Personal Service number | Name | Christian names | Remarks |
|---|---|---|---|---|---|
| BRONZE STAR | | | | | |
| WARRANT OFFICER CLASS II | | | | | |
| 40251 | 10.08.54 | 6/308 | RICHARDS | Edward J. | Britcom Workshops |

| London Gazette issue No. | London Gazette date | Personal Service number | Name | Christian names | Unit |
|---|---|---|---|---|---|
| AUSTRALIAN MILITARY PROVOST STAFF CORPS | | | | | |
| MBE: MEMBER OF THE ORDER OF THE BRITISH EMPIRE | | | | | |
| LIEUTENANT | | | | | |
| 40025 | 24.11.53 | 2/974 | SLATER | Harold Leslie | 1st Commonwealth Division Provost |
| MENTION IN DESPATCHES (FOR GALLANT AND DISTINGUISHED SERVICES) | | | | | |
| BRONZE OAK LEAF EMBLEM | | | | | |
| WARRANT OFFICER CLASS II | | | | | |
| 40025 | 24.11.53 | 1/221 | SAYEG | G.P. | 28th British Commonwealth Brigade |
| SERGEANT | | | | | |
| 40025 | 24.11.53 | 3/1993 | ALLWOOD | F.W. | 28th British Commonwealth Brigade |
| CORPORAL | | | | | |
| 40205 | 15.06.54 | 5/247 | DEIGHTON | N.A. | 1st Commonwealth Division Provost |

| London Gazette issue No. | London Gazette date | Personal Service number | Name | Christian names | Remarks |
|---|---|---|---|---|---|
| **AUSTRALIAN INTELLIGENCE CORPS** | | | | | |
| **MM: MILITARY MEDAL** | | | | | |
| 40205 | 15.06.54 | 3/833 | HARRIS | Alfred Martin | HQ 1st Commonwealth Division |
| **MENTION IN DESPATCHES (FOR GALLANT AND DISTINGUISHED SERVICES)** | | | | | |
| **BRONZE OAK LEAF EMBLEM** | | | | | |
| **CAPTAIN** | | | | | |
| 40205 | 15.06.54 | 2/37641 | LAMACRAFT | D.B. | 28th British Commonwealth Brigade |

| London Gazette issue No. | London Gazette date | Personal Service number | Name | Christian names | Remarks |
|---|---|---|---|---|---|
| ROYAL AUSTRALIAN ARMY CATERING CORPS | | | | | |
| MENTION IN DESPATCHES (FOR GALLANT AND DISTINGUISHED SERVICES) | | | | | |
| BRONZE OAK LEAF EMBLEM | | | | | |
| CORPORAL | | | | | |
| 40205 | 15.06.54 | 2/5028 | YOUNGMAN | D.A.W. | attached 2nd Bn R.A.R. |

# ROYAL AUSTRALIAN AIR FORCE

| London Gazette issue No. | London Gazette date | Personal Service number | Name | Christian names | Remarks |
|---|---|---|---|---|---|
| ROYAL AUSTRALIAN AIR FORCE | | | | | |
| DSO: DISTINGUISHED SERVICE ORDER | | | | | |
| WING COMMANDERS | | | | | |
| 40078 | 19.01.54 | 05833 | HUBBLE, AFC | John Wilkins | 77 Fighter/Interceptor Squadron |
| 39906 | 07.07.53 | 03104 | KINNINMONT, DFC | Jack Royston | 77 Fighter/Interceptor Squadron |
| 39594 | 11.07.52 | 04391 | SUSANS, DFC | Ronald Thomas | 77 Fighter/Interceptor Squadron |
| OBE: OFFICER OF THE ORDER OF THE BRITISH EMPIRE | | | | | |
| SQUADRON LEADER | | | | | |
| 39973 | 29.09.53 | 05856 | TONKIN | Ernest William | 91 (C) Wing |
| MBE: MEMBER OF THE ORDER OF THE BRITISH EMPIRE | | | | | |
| SQUADRON LEADERS | | | | | |
| | | 033035 | BARROW | S.J. | 77 Fighter/Interceptor Squadron |
| | | 03193 | HARDY | W.J. | 391 Squadron |
| FLIGHT LIEUTENANTS | | | | | |
| 39906 | 07.07.53 | 024256 | BROMHEAD | Stanley John | 391 Squadron |
| 39448 | 25.01.52 | 0510 | CASEY | Joseph Patrick | 391 Squadron |
| 39661 | 03.10.52 | 021998 | DAVIDSON | Ralph Russell | 77 Fighter/Interceptor Squadron |
| 39278 | 06.07.51 | 033072 | LYONS | Ian Alfred | 391 Squadron |
| 39731 | 30.12.52 | 023145 | SMITH | John Campbell | 491 Squadron |
| LIEUTENANT | | | | | |
| 40193 | 10.06.54 | 03498 | McCULLOUGH | Kenneth | 77 Fighter/Interceptor Squadron |
| WARRANT OFFICERS | | | | | |
| 39961 | 03.10.52 | A1188 | BEUTLER | John Lindsay | 491 Squadron |
| 39553 | 03.05.52 | A31144 | FLETCHER | Kenneth Maxwell | 491 Squadron |
| BAR TO DFC: DISTINGUISHED FLYING CROSS | | | | | |
| WING COMMANDER | | | | | |
| 39205 | 17.04.51 | 011315 | SPENCE, DFC | Louis Thomas | 77 Fighter/Interceptor Squadron listed: since deceased |
| SQUADRON LEADERS | | | | | |
| 39731 | 30.12.52 | 012289 | BENNETT, DFC | William Robert | 77 Fighter/Interceptor Squadron |
| 40078 | 19.01.54 | 021989 | GLASSOP, DFC | Ross Huggins | 77 Fighter/Interceptor Squadron |
| FLIGHT LIEUTENANTS | | | | | |
| 39278 | 06.07.51 | 04410 | OLORENSHAW, DFC | Ian Russell | 77 Fighter/Interceptor Squadron |
| 39731 | 30.12.52 | 022188 | RIVERS, DFC | Wallace Bolton | 77 Fighter/Interceptor Squadron |
| 39731 | 30.12.52 | 023682 | TURNER, DFC | Valton Leslie | 77 Fighter/Interceptor Squadron |

| London Gazette issue No. | London Gazette date | Personal Service number | Name | Christian names | Remarks |
|---|---|---|---|---|---|
| DFC: DISTINGUISHED FLYING CROSS | | | | | |
| SQUADRON LEADERS | | | | | |
| 39435 | 08.01.52 | 0383 | CRESSWELL | Richard | 77 Fighter/Interceptor Squardon |
| 39813 | 31.03.53 | 011391 | PARKER | Ian Stanley | 77 Fighter/Interceptor Squadron |
| 39448 | 25.01.52 | 021924 | WILSON | David Lindsay | 77 Fighter/Interceptor Squadron |
| FLIGHT LIEUTENANTS | | | | | |
| 39205 | 17.04.51 | 033119 | ADAMS, AFC | John Ihwin | 77 Fighter/Interceptor Squadron |
| 39287 | 06.07.51 | 033196 | BARNES | Frederick William | 77 Fighter/Interceptor Squadron |
| 39205 | 17.04.51 | 022040 | BRADFORD | Stuart | 77 Fighter/Interceptor Squadron |
| 39554 | 30.05.52 | 022094 | CANNON | Victor Benjamin | 77 Fighter/Interceptor Squadron |
| 39278 | 06.07.51 | 02179 | COBURN | Frank Ross | 77 Fighter/Interceptor Squadron |
| 39505 | 01.04.52 | 033109 | DAWSON | Ralph Leslie | 77 Fighter/Interceptor Squadron |
| 39891 | 16.06.53 | 035373 | GRIGGS, DFM | Frank Morton | 77 Fighter/Interceptor Squadron |
| 39709 | 02.12.52 | 033133 | GUY | Edgar Wesley | 77 Fighter/Interceptor Squadron |
| 39278 | 06.07.51 | 022048 | HARVEY | Gordon Ronald | 77 Fighter/Interceptor Squadron |
| 39973 | 29.09.53 | 033618 | HILL | Vincent Jerome | 77 Fighter/Interceptor Squadron |
| 39906 | 07.07.53 | 011371 | HURST | Douglas Charles | 77 Fighter/Interceptor Squadron |
| 39554 | 30.05.52 | 033135 | MARTIN | Keith Alexander | 77 Fighetr/Interceptor Squadron |
| | | 56713 | MELLERS | J. | 77 Fighter/Interceptor Squadron |
| 39731 | 30.12.52 | 033192 | MIDDLETON | Peter Montague | 77 Fighter/Interceptor Squadron |
| 39505 | 01.04.52 | 033188 | MURPHY | Cornelius Desmond | 77 Fighter/Interceptor Squadron |
| 39278 | 06.07.51 | 022053 | MURRAY | Colin John | 77 Fighter/Interceptor Squadron |
| 39205 | 17.04.51 | 033061 | NOBLE | Carlyle Richard | 77 Fighter/Interceptor Squadron |
| 39661 | 03.10.52 | 011561 | PURSSEY | Ian Goodwin Swan | 77 Fighter/Interceptor Squadron |
| 39802 | 17.03.53 | 05975 | RAMSAY | Eric George | 77 Fighter/Interceptor Squadron |
| | | 05974 | SHEARN | H.V. | 77 Fighter/Interceptor Squadron |
| 39892 | 19.06.53 | 05813 | SUGDEN | Christopher John | 77 Fighter/Interceptor Squadron |
| FLYING OFFICERS | | | | | |
| 39661 | 03.10.52 | 011402 | GOGERLY | Bruce | 77 Fighter/Interceptor Squadron |
| 39661 | 03.10.52 | 021993 | HAMILTON-FOSTER | Phillip Vincent | 77 Fighter/Interceptor Squadron |
| 39205 | 17.04.51 | 022102 | HORSMAN | William Charles | 77 Fighter/Interceptor Squadron |
| 39973 | 29.09.53 | 05827 | JOY | Brian | 77 Fighter/Interceptor Squadron |
| 39205 | 17.04.51 | 031479 | McLEOD | Kenneth Duncan | 77 Fighter/Interceptor Squadron |
| 39802 | 17.03.53 | 011417 | PHILP | Alan | 77 Fighter/Interceptor Squadron also listed as a Warrant Officer |
| 39448 | 25.01.52 | 05879 | READING | Leslie | 77 Fighter/Interceptor Squadron |
| 39906 | 07.07.53 | 022168 | TURNER | Avenal Richard | 77 Fighter/Interceptor Squadron |
| 39554 | 30.05.52 | 022058 | WITTMAN | Richard William | 77 Fighter/Interceptor Squadron |

| London Gazette issue No. | London Gazette date | Personal Service number | Name | Christian names | Remarks |
|---|---|---|---|---|---|
| DFC: DISTINGUISHED FLYING CROSS (continued) | | | | | |
| PILOT OFFICERS | | | | | |
| 39906 | 07.07.53 | 035498 | FOX | Raymond Francis | 77 Fighter/Interceptor Squadron |
| 39906 | 07.07.53 | 021144 | HUGHES | Henry Alfred | 77 Fighter/Interceptor Squadron |
| 39973 | 29.09.53 | 022790 | MURRAY, DFM | Kenneth James | 77 Fighter/Interceptor Squadron |
| 39973 | 29.09.53 | 011475 | SLATER | John William | 77 Fighter/Interceptor Squadron |
| 39278 | 06.07.51 | 033260 | TREBILCO | Raymond Edward | 77 Fighter/Interceptor Squadron |
| WARRANT OFFICERS | | | | | |
| 39435 | 08.01.52 | A33301 | HOWE | Charles Ronald Albert | 77 Fighter/Interceptor Squadron |
| 39505 | 01.04.52 | A11389 | HUNT | Robert Charles Arthur | 77 Fighter/Interceptor Squadron |
| 39278 | 06.07.51 | A11412 | MICHELSON | William Stapylton | 77 Fighter/Interceptor Squadron |
| 39278 | 06.07.51 | A22108 | RIVERS | Wallace Bolton | 77 Fighter/Interceptor Squadron also listed as A/Flight Lieutenant |
| 39278 | 06.07.51 | A33213 | WILLIAMSON | Stanley William | 77 Fighter/Interceptor Squadron |
| BAR TO AFC: AIR FORCE CROSS | | | | | |
| SQUADRON LEADER | | | | | |
| 39906 | 07.07.53 | 011326 | MURDOCH, AFC | Rodney Sinclair | 30 (TPT) Unit |
| AFC: AIR FORCE CROSS | | | | | |
| SQUADRON LEADERS | | | | | |
| | | 031718 | FIVASH | C.F. | 77 Fighter/Interceptor Squadron |
| 39505 | 01.04.52 | 03486 | GERBER | John Ewen | 30 COMM Unit |
| FLIGHT LIEUTENANTS | | | | | |
| 39802 | 17.03.53 | 011344 | ADDISON | Warwick | 30 (TPT) Unit |
| 39554 | 30.05.52 | 033106 | BARKLA | Franklyn Thomas | 30 (TPT) Unit |
| 39973 | 29.09.53 | 022020 | DANIEL | Ronald William | 77 Fighter/Interceptor Squadron |
| 39278 | 06.07.51 | 05820 | ELIOT | Noel Stirling | 30 COMM Unit |
| 39278 | 06.07.51 | 05836 | HITCHINS | David Wilson | 30 COMM Unit (also listed: HUTCHINS) |
| 39802 | 17.03.53 | 021969 | LYNCH | James Joseph | 30 (TPT) Unit |
| 39661 | 03.10.52 | 011330 | McDONALD | Joseph Kevin | 30 (TPT) Unit |
| 40078 | 19.01.54 | 031492 | MULLER | Frederick Keith | 36 (TPT) Unit |
| 39731 | 30.12.52 | 021999 | PARKER | Bernard | 30 (TPT) Unit |
| 30554 | 30.05.52 | 022043 | THOMAS | John Burgoyne | 30 (TPT) Unit |
| FLYING OFFICER | | | | | |
| | | A33201 | THORNTON, AFM | Geoffrey | 77 Fighter/Interceptor Squadron |
| WARRANT OFFICER | | | | | |
| 39435 | 08.01.52 | A22055 | RYAN | William Patrick | 30 (TPT) Unit |

| London Gazette issue No. | London Gazette date | Personal Service number | Name | Christian Names | Remarks |
|---|---|---|---|---|---|
| DFM: DISTINGUISHED FLYING MEDAL | | | | | |
| PILOT I (WARRANT OFFICERS) | | | | | |
| 39205 | 17.04.51 | A22103 | NICHOLLS | Brian Frederick Stanley | 77 Fighter/Interceptor Squadron |
| 39205 | 17.04.51 | A33201 | THORNTON, AFM | Geoffrey | 77 Fighter/Interceptor Squadron |
| FLIGHT-SERGEANTS | | | | | |
| 39505 | 01.04.52 | A32245 | BESSELL | Henry William | 77 Fighter/Interceptor Squadron |
| 39505 | 01.04.52 | A33283 | MEGGS | Keith Raymond | 77 Fighter/Interceptor Squadron |
| 39802 | 17.03.53 | A22110 | STONEY | Alfred Thomas | 77 Fighter/Interceptor Squadron |
| SERGEANTS | | | | | |
| 39906 | 07.07.53 | A34233 | ARMSTRONG | Tony Charles | 77 Fighter/Interceptor Squadron |
| 39553 | 30.05.52 | A5895 | COLEBROOK | Maxwell Edwin | 77 Fighter/Interceptor Squadron |
| 39435 | 18.01.52 | A4443 | FOSTER | Kevin Henry | 77 Fighter/Interceptor Squadron |
| 40078 | 19.01.54 | A35096 | JONES | Henry Edward | 77 Fighter/Interceptor Squadron |
| 39802 | 17.03.53 | A33830 | KING | Colin George | 77 Fighter/Interceptor Squadron |
| 39906 | 07.07.53 | A34254 | LUSHEY | Geoffrey Warren | 77 Fighter/Interceptor Squadron |
| 39661 | 03.10.52 | A22790 | MURRAY | Kenneth James | 77 Fighter Interceptor Squadron |
| 39661 | 03.10.52 | A22193 | MYERS | Edward John | 77 Fighter/Interceptor Squadron |
| 39661 | 03.10.52 | A33630 | OBORN | Ralph Victor | 77 Fighter/Interceptor Squadron |
| 39731 | 30.12.52 | A33838 | PARKER | John Norris | 77 Fighter/Interceptor Squadron |
| 39435 | 08.01.52 | A33287 | SLY | Cecil | 77 Fighter/Interceptor Squadron |
| 39554 | 30.05.52 | A33635 | STRAWBRIDGE | Robert Andrew | 77 Fighter/Interceptor Squadron |
| 39731 | 30.12.52 | A21121 | TOWNER | Kenneth Graham | 77 Fighter/Interceptor Squadron |
| BEM: BRITISH EMPIRE MEDAL | | | | | |
| FLIGHT SERGEANTS | | | | | |
| | | A2317 | CHAFFEY | R.E. | 77 Fighter/Interceptor Squadron |
| | | A2171 | FREEMAN | M.P. | 77 Fighter/Interceptor Squadron |
| | | A2243 | MARTIN | R.J. | 391 Squadron |
| MENTION IN DESPATCHES (FOR GALLANT AND DISTINGUISHED SERVICES) | | | | | |
| BRONZE OAK LEAF EMBLEM | | | | | |
| FLIGHT-LIEUTENANTS | | | | | |
| 39731 | 30.12.52 | 024256 | BROMHEAD | Stanley John | 391 Squadron |
| 39661 | 30.10.51 | 033097 | CADAN | Lawrence Leslie | 77 Fighter/Interceptor Squadron |
| 39973 | 29.09.53 | 022017 | GODFREY | Kenneth Henry | 391 Squadron |
| 39802 | 17.03.53 | 03286 | JOHNSTON | Henry Eric | 77 Fighter/Interceptor Squadron |
| 39278 | 06.07.51 | 031717 | McCROHAN | Thomas Eugene | 77 Fighter/Interceptor Squadron also listed as McCROHAM |
| 39906 | 07.07.53 | 03498 | McCULLOUGH | Kenneth | 77 Fighter/Interceptor Squadron |

| London Gazette issue No. | London Gazette date | Personal Service number | Name | Christian names | Remarks |
|---|---|---|---|---|---|
| MENTION IN DESPATCHES (FOR GALLANT AND DISTINGUISHED SERVICES) | | | | | |
| BRONZE OAK LEAF EMBLEM | | | | | |
| FLIGHT-LIEUTENANTS (continued) | | | | | |
| 39906 | 07.07.53 | 036133 | MORRISON | James McLean | 77 Fighter/Interceptor Squadron |
| 39278 | 06.07.51 | 033171 | MURPHY | Thomas William | 77 Fighter/Interceptor Squadron |
| 39661 | 03.10.52 | 033201 | THORNTON, DFM, AFM | Geoffrey | 77 Fighter/Interceptor Squadron |
| FLYING OFFICERS | | | | | |
| 39278 | 06.07.51 | 022136 | FLEMMING | James Hilary | 77 Fighter/Interceptor Squadron |
| 39278 | 06.07.51 | 022133 | GARROWAY | William Maxwell | 77 Fighter/Interceptor Squadron |
| 39906 | 07.07.53 | 033271 | GREEN | Randall Robert | 77 Fighter/Interceptor Squadron |
| 39448 | 25.01.52 | 021993 | HAMILTON-FOSTER | Philip Vincent | 77 Fighter/Interceptor Squadron |
| 39806 | 17.03.53 | 04424 | PRAYNE | Ross McRostie | 77 Fighter/Interceptor Squadron also listed as FRAYNE and for U.S. Air Medal award LG 39999 |
| PILOT OFFICERS | | | | | |
| 39906 | 07.07.53 | 032529 | BOORD | Griffith Alexander | 77 Fighter/Interceptor Squadron |
| 40169 | 07.05.54 | 035083 | COLLINS | Geoffrey Anthony | 77 Fighter/Interceptor Squadron |
| 39278 | 06.07.51 | 022222 | COTTEE | Milton James | 77 Fighter/Interceptor Squadron |
| 40078 | 19.01.54 | 033089 | GILMOUR | John Leslie | 77 Fighter/Interceptor Squadron |
| 40365 | 31.12.54 | 05309 | HALLEY | John Beverley | 77 Fighter/Interceptor Squadron listed: since deceased |
| 39906 | 07.07.53 | 032533 | HOWARD | Hugh Bryan | 77 Fighter/Interceptor Squadron |
| 39435 | 08.01.52 | 04218 | KLAFFER | Lyall Robert | 77 Fighter/Interceptor Squadron |
| 39973 | 29.09.53 | 033284 | MULLEN | Bernard Thomas | 77 Fighter/Interceptor Squadron |
| 39906 | 07.07.53 | 032542 | SIMMONDS | William Henry | 77 Fighter/Interceptor Squadron |
| 39553 | 30.05.52 | 033260 | TREBILCO, DFC | Raymond Edward | 77 Fighter/Interceptor Squadron |
| WARRANT OFFICERS | | | | | |
| 39278 | 06.07.51 | A22086 | BRACKENREG | Roydon Owen Leonard | 77 Fighter/Interceptor Squadron |
| 39278 | 06.07.51 | A3960 | CARLSON | William John | 77 Fighter/Interceptor Squadron |
| 39278 | 06.07.51 | A33027 | FAIRWEATHER | Robert Forrester | 77 Fighter/Interceptor Squadron |
| 39731 | 30.12.52 | A31102 | FIELDS | James | 491 Squadron |
| 39661 | 03.10.52 | A11424 | HILL | Keith Gordon | 77 Fighter/Interceptor Squadron |
| 39906 | 07.09.53 | A31219 | HOEY | George Joseph | ← Unit not listed |
| 39278 | 06.07.51 | A3823 | McEACHERN | Alexander William | 391 Squadron |
| 39661 | 03.10.52 | A3901 | STAPLETON | Thomas Lawrence | 491 Squadron |
| 39435 | 08.01.52 | A22168 | TURNER | Avenal Richard | 77 Fighter/Interceptor Squadron |
| 39892 | 19.06.53 | A39731 | TURNER | Robert | 391 Squadron Service number also listed as A4434 |
| 39553 | 30.05.52 | A31516 | WEBSTER | James Bertinias | 491 Squadron |

| London Gazette issue No. | London Gazette date | Personal Service number | Name | Christian names | Remarks |
|---|---|---|---|---|---|
| MENTION IN DESPATCHES (FOR GALLANT AND DISTINGUISHED SERVICES)(continued) | | | | | |
| BRONZE OAK LEAF EMBLEM | | | | | |
| FLIGHT-SERGEANTS | | | | | |
| 39661 | 03.10.52 | A22140 | AVERY, DFM | Allan James | 77 Fighter/Interceptor Squadron |
| 39278 | 06.07.51 | A413 | BRAMLEY | Phillip Richard | 77 Fighter/Interceptor Squadron |
| 39278 | 06.07.51 | A2122 | BREEN | George James Francis | 391 Squadron |
| 39278 | 06.07.51 | A31296 | FALCONER | Hugh George | 391 Squadron |
| 39278 | 06.07.51 | A31509 | FAWCETT | Henry Kenneth Westmore | 491 Squadron(also listed as Sergeant) |
| 39278 | 06.07.51 | A31409 | FREEMAN | John Joseph | 77 Fighter/Interceptor Squadron also listed as Sergeant |
| 39906 | 07.07.53 | A32378 | JANSON | Kenneth Robert | 77 Fighter/Interceptor Squadron |
| 40078 | 19.01.54 | A31074 | JOHNSTON | George | 391 Squadron |
| 39973 | 29.09.53 | A31415 | KAY | John Ezra | 77 Fighter/Interceptor Squadron |
| 39278 | 06.07.51 | A2369 | McALEER | Raymond Charles | 391 Squadron(also listed as Sergeant) |
| 39891 | 19.06.53 | A21 | MURRAY | Gordon Douglas | 77 Fighter/Interceptor Squadron |
| 39435 | 08.01.52 | A31485 | SANDY | Reginald James | 391 Squadron |
| 39435 | 08.01.52 | A450 | TITLEY | Geoffrey | 77 Fighter/Interceptor Squadron |
| | | A1144 | VICTOR | John Nugent | 77 Fighter/Interceptor Squadron |
| SERGEANTS | | | | | |
| 39504 | 01.04.52 | A141 | ALLEN | Paul Alexander | 77 Fighter/Interceptor Squadron |
| 39906 | 07.07.53 | A33621 | BLACKWELL | Francis Robert | 77 Fighter/Interceptor Squadron |
| 39906 | 07.07.53 | A31614 | CAHILL | John Vincent | 77 Fighter/Interceptor Squadron |
| 40169 | 07.05.54 | A34495 | COLLINGS | Billie Hicks | 77 Fighter/Interceptor Squadron |
| 39973 | 29.09.53 | A31848 | CORDWELL | Thomas James | 77 Fighter/Interceptor Squadron |
| 39448 | 23.01.52 | A562 | CORRY | Henry Kingsley | 391 Squadron |
| 39802 | 17.03.53 | A549 | DUGGAN | John Patrick Russell | 391 Squadron |
| 39802 | 17.03.53 | A22613 | EVANS | John Alexander | 77 Fighter/Interceptor Squadron |
| 40078 | 19.01.54 | A35089 | GILMOUR | John Leslie | 77 Fighter/Interceptor Squadron also listed as Pilot Officer |
| 39553 | 30.05.52 | A33231 | GIULIERI | Reginald | 491 Squadron |
| 40078 | 19.01.54 | A223 | GRAHAM | Robert Alan | 77 Fighter/Interceptor Squadron |
| 39802 | 17.03.53 | A33837 | GRAY | Paul Turton | 91 (C) Wing Service number also listed as A35132 |
| 40078 | 19.01.54 | A35090 | HALE | George Spaulding | 77 Fighter/Interceptor Squadron |
| 39802 | 17.03.53 | A4125 | HILLMAN | Ronald Leslie | 391 Squadron |
| 39554 | 30.05.52 | A2658 | JARRETT | John Bernard | 421 (Maintenance) Squadron also listed as 491 Squadron |
| 39802 | 17.03.53 | A33628 | KITCHENSIDE | James Charles | 77 Fighter/Interceptor Squadron |
| 39534 | 08.01.52 | A31059 | MALIN | Alic Hilton | 491 Squadron |

| London Gazette issue No. | London Gazette date | Personal Service number | Name | Christian names | Remarks |
|---|---|---|---|---|---|
| MENTION IN DESPATCHES (FOR GALLANT AND DISTINGUISHED SERVICES) (continued) | | | | | |
| BRONZE OAK LEAF EMBLEM | | | | | |
| SERGEANTS (continued) | | | | | |
| 39554 | 30.05.52 | A2265 | MULCAHY | Phillip James | 77 Fighter/Interceptor Squadron |
| 39802 | 17.03.53 | A33837 | OUTHWAITE | Maxwell Allan | 77 Fighter/Interceptor Squadron |
| 39435 | 08.01.52 | A2482 | PIGGOTT | Kenneth Alfred | 491 Squadron |
| 39728 | 06.07.51 | A242 | RAVENSCROFT | William Linton | 391 Squadron |
| 39278 | 06.07.51 | A11436 | REBURN | Neville | 391 Squadron |
| 39553 | 30.05.52 | A11436 | REBURN | Neville | 391 Squadron |
| CORPORALS | | | | | |
| 39448 | 05.01.52 | A2785 | MACKENZIE | William Alexander | 30 COMM Unit |
| 39435 | 08.01.52 | A2550 | McMILLAN | Keith William | 77 Fighter/Interceptor Squadron |
| 39435 | 08.01.52 | A1274 | MADSEN | Neils Francis | 491 Squadron |
| 39661 | 03.10.52 | A11060 | MELLICAN | Kevin Patrick | 491 Squadron |
| 40169 | 07.05.54 | A21293 | MOONEY | Bryan Timothy | 77 Fighter/Interceptor Squadron |
| 39554 | 30.05.52 | A33256 | MORGAN | John Murray | 491 Squadron |
| 39448 | 25.01.52 | A31800 | MORRISON | Lionel | 30 COMM Unit |
| 39435 | 08.01.52 | A31914 | OEHM | Victor Herbert | 391 Squadron |
| 39802 | 17.03.53 | A31781 | PAYNE | Ronald Roy | 77 Fighter/Interceptor Squadron |
| 39802 | 17.03.53 | A1418 | RINEHART | Patrick Frederick | 77 Fighter/Interceptor Squadron |
| 39973 | 29.09.53 | A31648 | SANFEAD | Thomas Clifford | 77 Fighter/Interceptor Squadron |
| 39278 | 06.07.51 | A31908 | SLATER | Thomas William Wood | 491 Squadron |
| 39802 | 17.03.53 | A1249 | SULLIVAN | Noel Joseph | 77 Fighter/Interceptor Squadron |
| 39448 | 25.01.52 | A31832 | SUNBERG | Gerald Joseph | 491 Squadron |
| 39802 | 17.03.53 | A1713 | SYMES | Gordon Stanley | 77 Fighter/Interceptor Squadron |
| 39435 | 08.01.52 | A22204 | THORPE | Richard Mervyn | 391 Squadron also listed as Flight-Sergeant |
| 39661 | 03.10.52 | A1363 | TUCKER | Malcolm Charles | 77 Fighter/Interceptor Squadron |
| 39973 | 29.09.53 | A34192 | VERDON | John Francis | 77 Fighter/Interceptor Squadron |
| 39504 | 01.04.52 | A22062 | WHITBREAD | Gregory James | 391 Squadron |
| 39435 | 08.01.52 | A2886 | WILLIAMS | Ronald Dennis | 391 Squadron |
| LEADING AIRCRAFTMEN | | | | | |
| 39584 | 01.04.52 | A5378 | AMBROSE | Everett Arthur | 30 COMM Unit |
| 39584 | 01.04.52 | A21126 | BEGGS | Ronald Richardson | 77 Fighter/Interceptor Squadron |
| 39731 | 30.12.52 | A21324 | BLACKLEY | Keith Andrew | 77 Fighter/Interceptor Squadron |
| 39584 | 01.04.52 | A11085 | BRADY | John Francis | 491 Squadron |
| 39802 | 17.03.53 | A1917 | BREEZE | Robert Stanley | 77 Fighter/Interceptor Squadron |
| 39435 | 08.01.52 | A2597 | BUTLER | William Thomas | 77 Fighter/Interceptor Squadron |
| 40169 | 07.05.54 | A23378 | CALLAGHAN | Lawrence | 77 Fighter/Interceptor Squadron |

| London Gazette issue No. | London Gazette date | Personal Service number | Name | Christian names | Remarks |
|---|---|---|---|---|---|
| MENTION IN DESPATCHES (FOR GALLANT AND DISTINGUISHED SERVICES) (continued) | | | | | |
| BRONZE OAK LEAF EMBLEM | | | | | |
| LEADING AIRCRAFTMEN (continued) | | | | | |
| 39802 | 17.03.53 | A23148 | CLARK | Noel John | 77 Fighter/Interceptor Squadron |
| 39731 | 30.12.52 | A24085 | COLLINS | William George | 491 Squadron |
| 39584 | 01.04.52 | A145 | CRONIN | Lloyd | 91 (C) Wing |
| 39435 | 08.01.52 | A31835 | DALTON | Sydney | 491 Squadron |
| 39973 | 29.09.53 | A33527 | DAWSON | Stanley | 77 Fighter/Interceptor Squadron |
| 39435 | 08.01.52 | A22602 | DAWSON | Terence Darrell | 491 Squadron |
| 39802 | 17.03.53 | A11097 | DOYLE | Leslie William | 77 Fighter/Interceptor Squadron |
| 39584 | 01.04.52 | A1647 | EINAM | Keith Archibald | 77 Fighter/Interceptor Squadron |
| 39553 | 30.05.52 | A31905 | EVANS | Keith Ernest | 77 Fighter/Interceptor Squadron |
| 40169 | 07.05.54 | A33373 | JACKSON | Stanley James | 77 Fighter/Interceptor Squadron |
| 39553 | 30.05.52 | A1205 | JOHNSTON | Gordon Douglas | 77 Fighter/Interceptor Squadron |
| 40078 | 19.01.54 | A24397 | JONES | Brian Mohr | 391 Squadron |
| 39731 | 30.12.52 | A11762 | KELLY | Gordon Duke | 77 Fighter/Interceptor Squadron |
| 39584 | 01.04.52 | A33398 | KERR | Gilbert George | 77 Fighter/Interceptor Squadron |
| 39661 | 03.10.52 | A23039 | PEARSON | Ronald James | 391 Squadron |
| 39584 | 01.04.52 | A22701 | PHILBROOK | Robert Arthur | 77 Fighter/Interceptor Squadron |
| 39731 | 30.12.52 | A12217 | RICHARDSON | Alexander | 391 Squadron |
| 40169 | 07.05.54 | A22460 | ROACH | Frederick William | 77 Fighter/Interceptor Squadron |
| 39553 | 30.05.52 | A34194 | SANDMAN | Mervyn Hedley | 391 Squadron |
| 39802 | 17.03.53 | A1663 | SHEPHARD | Earl Bruce | 77 Fighter/Interceptor Squadron |
| 39435 | 08.01.52 | A32611 | SILLING | Charles Malcolm | 77 Fighter/Interceptor Squadron |
| 39731 | 30.12.52 | A23158 | THOMAS | James Eric | 491 Squadron |
| 39731 | 30.12.52 | A6170 | TURNER | Brian Rex | 391 Squadron |
| 39435 | 08.01.52 | A5206 | WALTERS | Stanley Lionel | 491 Squadron |
| 39731 | 30.12.52 | A21634 | WATTERSON | Vivian Norman | 77 Fighter/Interceptor Squadron (Service number also listed as A1986) |
| 39661 | 03.10.52 | A1936 | WELBURN | Heber | 391 Squadron also listed as WELLBURN |
| 40078 | 19.01.54 | A22325 | WHITE | Brian Francis | 77 Fighter/Interceptor Squadron |
| 39731 | 30.12.52 | A21634 | WILLIAMS | Walter John Edgar | 491 Squadron |
| KING'S COMMENDATION (FOR VALUABLE SERVICES IN THE AIR IN CONNECTION WITH KOREAN WAR OPERATIONS) | | | | | |
| BRONZE OAK LEAF EMBLEM | | | | | |
| FLIGHT-LIEUTENANT | | | | | |
| 39435 | 08.01.52 | 03541 | TAFE | Alfred Frederick | 30 COMM Unit |

| London Gazette issue No. | London Gazette date | Personal Service number | Name | Christian names | Remarks |
|---|---|---|---|---|---|
| KING'S COMMENDATION (FOR VALUABLE SERVICES IN THE AIR IN CONNECTION WITH KOREAN WAR OPERATIONS) (continued) | | | | | |
| BRONZE OAK LEAF EMBLEM | | | | | |
| FLYING OFFICERS | | | | | |
| 39435 | 08.01.52 | 022191 | BRAY | Robert Derik | 39 COMM Unit |
| 39435 | 08.01.52 | 022071 | ROBERTS | Alick George | 30 COMM Unit |
| WARRANT OFFICERS | | | | | |
| 39448 | 25.01.52 | A22190 | BURNS | Harold Douglas Haigh | 30 COMM Unit |
| 39448 | 25.01.52 | A22118 | HARDMAN | Stephen Phillip | not listed |
| 39435 | 08.01.52 | A22126 | LANG | Neil Raymond | 30 COMM Unit |
| KING'S COMMENDATION (FOR DISTINGUISHED AND VALUABLE SERVICES IN THE AIR) | | | | | |
| BRONZE OAK LEAF EMBLEM | | | | | |
| FLIGHT-LIEUTENANT | | | | | |
| 39278 | 06.07.51 | 011330 | McDONALD | Joseph Kevin | 30 COMM Unit |
| SERGEANT | | | | | |
| 39278 | 06.07.51 | A33300 | HURLEY | Henry James | 30 COMM Unit |
| QUEEN'S COMMENDATION (FOR DISTINGUISHED AND VALUABLE SERVICES IN THE AIR IN KOREA) | | | | | |
| BRONZE OAK LEAF EMBLEM | | | | | |
| FLIGHT-LIEUTENANTS | | | | | |
| 39661 | 03.10.52 | 035366 | BREMNER | Colin David | 30 COMM Unit |
| 39554 | 30.05.52 | 022037 | DONNELLY | Clarence Armfield | 30 COMM Unit |
| 39661 | 03.10.52 | 023961 | KEITH | Jack Watson | 30 COMM Unit |
| 40169 | 07.05.54 | 022074 | McLEOD | Thomas Lindsay | 36 (TPT) Squadron |
| FLYING OFFICER | | | | | |
| 39661 | 03.10.52 | 023624 | SPINKS | William Thomas | 30 COMM Unit |
| QUEEN'S COMMENDATION (FOR VALUABLE SERVICES IN THE AIR) | | | | | |
| BRONZE OAK LEAF EMBLEM | | | | | |
| FLIGHT-LIEUTENANT | | | | | |
| 39973 | 29.09.53 | 033221 | PRETTY | Walter Ivan | 36 (TPT) Squadron (also listed in LG 39265 as Christian name IRAN, only) |
| FLIGHT-SERGEANTS | | | | | |
| 39973 | 29.09.53 | A33638 | FITZGERALD | Martin Joseph | 30 (TPT) Squadron |
| 39973 | 29.09.53 | A22206 | MUTAGH | Leon Francis | 36 (TPT) Squadron |

| London Gazette issue No. | London Gazette date | Personal Service number | Name | Christian names | Remarks |
|---|---|---|---|---|---|

## AWARDS CONFERRED BY THE PRESIDENT OF THE UNITED STATES OF AMERICA

| London Gazette issue No. | London Gazette date | Personal Service number | Name | Christian names | Remarks |
|---|---|---|---|---|---|
| LEGION OF MERIT, DEGREE OF OFFICER | | | | | |
| WING COMMANDER | | | | | |
| 39265 | 19.06.51 | 011315 | SPENCE, DFC | Louis Thomas | 77 Fighter/Interceptor Squadron (listed: since deceased) |
| LEGION OF MERIT, DEGREE OF LEGIONNAIRE | | | | | |
| GROUP CAPTAINS | | | | | |
| | | 0316 | CARR | A.G. | 91 (C) Wing |
| | | 0335 | CHAPMAN | D.R. | 91 (C) Wing |
| DISTINGUISHED FLYING CROSS | | | | | |
| WING COMMANDER | | | | | |
| 39999 | 27.10.53 | 04391 | SUSANS, DSO, DFC | Ronald Thomas | 77 Fighter/Interceptor Squadron |
| SQUADRON LEADERS | | | | | |
| 39999 | 27.10.53 | 012289 | BENNETT, DFC | William Robert | 77 Fighter/Interceptor Squadron |
| 39285 | 13.07.51 | 0383 | CRESSWELL | Richard | 77 Fighter/Interceptor Squadron |
| 39531 | 02.05.52 | 04410 | OLORENSHAW, DFC | Ian Russell | 77 Fighter/Interceptor Squadron |
| FLIGHT-LIEUTENANTS | | | | | |
| 39265 | 22.06.51 | 022046 | HARVEY | Gordon Ronald | 77 Fighter/Interceptor Squadron (listed: missing) |
| 39531 | 02.05.52 | 033171 | MURPHY | Thomas William | 77 Fighter/Interceptor Squadron |
| FLYING OFFICERS | | | | | |
| 39531 | 02.05.52 | 022086 | BRACKENREG | Owen Leonard | 77 Fighter/Interceptor Squadron (also listed as Warrant Officer Service number A22086) |
| 39531 | 02.05.52 | 022136 | FLEMING | James Hilary | 77 Fighter/Interceptor Squadron (also listed LG39278 for MID award as FLEMMING) |
| 39531 | 02.05.52 | 022133 | GARROWAY | William Maxwell | 77 Fighter/Interceptor Squadron |
| | | 011402 | GOGERLY | Bruce | 77 Fighter/Interceptor Squadron |
| 39999 | 27.10.53 | 022108 | RIVERS, DFC | Wallace Bolton | 77 Fighter/Interceptor Squadron |
| PILOT OFFICERS | | | | | |
| 39265 | 22.06.51 | 021261 | ELLIS | Donald Campbell | 77 Fighter/Interceptor Squadron (listed: missing) |
| 39265 | 22.06.51 | 04218 | KLAFFER | Lyall Robert | 77 Fighter/Interceptor Squadron |

| London Gazette issue No. | London Gazette date | Personal Service number | Name | Christian names | Remarks |
|---|---|---|---|---|---|
| DISTINGUISHED FLYING CROSS (USA) | | | | | |
| PILOT OFFICERS (continued) | | | | | |
| 39999 | 27.10.53 | 033630 | OBORN, DFM | Ralph Victor | 77 Fighter/Interceptor Squadron (also listed as A33630 SERGEANT for DFM award LG39661 03.10.52) |
| 39265 | 22.06.51 | 0647 | STEPHENS | Geoffrey Ingram | 77 Fighter/Interceptor Squadron |
| WARRANT OFFICERS | | | | | |
| | | A11424 | HILL | Keith Gordon | 77 Fighter/Interceptor Squadron |
| 39265 | 22.06.51 | A22168 | TURNER | Avenal Richard | 77 Fighter/Interceptor Squadron (also listed as Flying Officer) |
| SERGEANTS | | | | | |
| 39531 | 02.05.52 | A33217 | ROYAL | Kenneth Edward | 77 Fighter/Interceptor Squadron (listed: since deceased) |
| 39285 | 13.07.51 | A22110 | STONEY | Alfred Thomas | 77 Fighter/Interceptor Squadron |
| BRONZE STAR | | | | | |
| FLIGHT-LIEUTENANT | | | | | |
| 39531 | 02.05.52 | 033072 | LYONS, MBE | Ian Alfred | 77 Fighter/Interceptor Squadron |
| AIR MEDAL | | | | | |
| WING COMMANDERS | | | | | |
| | | 03104 | KINNINMONT | Jack Royston | 77 Fighter/Interceptor Squadron |
| 39265 | 22.06.51 | 011315 | SPENCE, DFC | Louis Thomas | 77 Fighter/Interceptor Squadron (listed: since deceased) |
| 39999 | 27.10.53 | 04391 | SUSANS, DSO, DFC | Ronald Thomas | 77 Fighter/Interceptor Squadron |
| 39531 | 02.05.52 | 021924 | WILSON, DFC | David Lindsay | 77 Fighter/Interceptor Squadron (also listed as Squadron Leader) |
| SQUADRON LEADERS | | | | | |
| 39999 | 27.10.53 | 012289 | BENNETT, DFC | William Robert | 77 Fighter/Interceptor Squadron |
| 39999 | 27.10.53 | 022094 | CANNON, DFC | Victor Benjamin | 77 Fighter/Interceptor Squadron |
| 39265 | 22.06.51 | 0383 | CRESSWELL | Richard | 77 Fighter/Interceptor Squadron |
| 39999 | 27.10.53 | 033133 | GUY, DFC | Edgar Wesley | 77 Fighter/Interceptor Squadron |
| | | 011371 | HURST | Douglas Charles | 77 Fighter/Interceptor Squadron |
| 39999 | 27.10.53 | 05806 | O'DONNELL, DFC | John Joseph | 77 Fighter/Interceptor Squadron |
| | | 011391 | PARKER | Ian Stanley | 77 Fighter/Interceptor Squadron (also listed as FLIGHT-LIEUTENANT) |
| 39999 | 27.10.53 | 011384 | THOMAS, DFC | Cedric George | 77 Fighter/Interceptor Squadron (also listed as FLIGHT-LIEUTENANT) |

| London Gazette issue No. | London Gazette date | Personal Service number | Name | Christian names | Remarks |
|---|---|---|---|---|---|
| AIR MEDAL (continued) | | | | | |
| FLIGHT-LIEUTENANTS | | | | | |
| 39265 | 22.06.51 | 033119 | ADAMS, DFC, AFC | John Ihwin | 77 Fighter/Interceptor Squadron |
| 39265 | 22.06.51 | 033196 | BARNES | Frederick Williams | 77 Fighter/Interceptor Squadron |
| | | 032529 | BOORD | Griffith Alexander | 77 Fighter/Interceptor Squadron |
| 39265 | 22.06.51 | 22040 | BRADFORD, DFC | Stuart | 77 Fighter/Interceptor Squadron |
| 39999 | 27.10.53 | 022087 | BROWN | Leon Barry | 77 Fighter/Interceptor Squadron |
| 39999 | 27.10.53 | 023665 | BROWNE-GAYLORD | Mark Astil Baren Henry Aytack | 77 Fighter/Interceptor Squadron (listed: missing) |
| 39999 | 27.10.53 | 033097 | CADAN | Lawrence Leslie | 77 Fighter/Interceptor Squadron |
| 39265 | 22.06.51 | 02179 | COBURN | Frank Ross | 77 Fighter/Interceptor Squadron |
| 39265 | 22.06.51 | 033109 | DAWSON | Ralph Leslie | 77 Fighter/Interceptor Squadron |
| 39999 | 27.10.53 | 022176 | GRAY | William Victor | 77 Fighter/Interceptor Squadron (listed: since deceased) |
| | | 033271 | GREEN | Randall Robert | 77 Fighter/Interceptor Squadron |
| 39999 | 27.10.53 | 022100 | HANNAN | John Thomas | 77 Fighter/Interceptor Squadron (listed:missing) |
| 39265 | 22.06.51 | 022048 | HARVEY | Gordon Ronald | 77 Fighter/Interceptor Squadron (listed:missing) |
| 39265 | 22.06.51 | 022102 | HORSMAN, DFC | William Charles | 77 Fighter/Interceptor Squadron |
| | | 021144 | HUGHES | Henry Alfred | 77 Fighter/Interceptor Squadron |
| 39265 | 22.06.51 | 023170 | KIRKPATRICK | Craig | 77 Fighter/Interceptor Squadron (listed: since deceased) |
| 39265 | 22.06.51 | 031717 | McCROHAN | Thomas Eugene | 77 Fighter/Interceptor Squadron |
| 39265 | 22.06.51 | 031479 | McLEOD, DFC | Kenneth Duncan | 77 Fighter/Interceptor Squadron |
| 39999 | 27.10.53 | 033134 | MARTIN, DFC | Keith Alexander | 77 Fighter/Interceptor Squadron |
| 39999 | 27.10.53 | 033192 | MIDDLETON, DFC | Peter Montague | 77 Fighter/Interceptor Squadron |
| 39265 | 22.06.51 | 033188 | MURPHY | Cornelius Desmond | 77 Fighter/Interceptor Squadron |
| 39265 | 22.06.51 | 033171 | MURPHY | Thomas William | 77 Fighter/Interceptor Squadron |
| 39265 | 22.06.51 | 022053 | MURRAY | John Collin | 77 Fighter/Interceptor Squadron |
| 39265 | 22.06.51 | 033061 | NOBLE, DFC | Carlyle Richard | 77 Fighter/Interceptor Squadron |
| 39265 | 22.06.51 | 04410 | OLORENSHAW, DFC | Ian Russell | 77 Fighter/Interceptor Squadron |
| 39265 | 22.06.51 | 033221 | PRETTY | Iran | 77 Fighter/Interceptor Squadron (also listed in LG 39973 as Christian names WALTER IVAN for Queen's Commendation) |
| 39999 | 27.10.53 | 011561 | PURSSEY, DFC | Ian Goodwin Swan | 77 Fighter/Interceptor Squadron |
| 39999 | 27.10.53 | 05975 | RAMSAY, DFC | Eric George | 77 Fighter/Interceptor Squadron |
| 39265 | 22.06.51 | 0428 | TAPLIN | Colin Charles | 77 Fighter/Interceptor Squadron |

| London Gazette issue No. | London Gazette date | Personal Service number | Name | Christian name | Remarks |
|---|---|---|---|---|---|
| AIR MEDAL (continued) | | | | | |
| FLIGHT-LIEUTENANTS | | | | | |
| 39999 | 27.10.53 | 05811 | TAYLOR | Raymond Alexander Ellwood | 77 Fighter/Interceptor Squadron |
| 39999 | 27.10.53 | 023862 | TURNER, DFC | Valton Leslie John | 77 Fighter/Interceptor Squadron |
| FLYING OFFICERS | | | | | |
| 39999 | 27.10.53 | 031569 | BLIGHT | Kenneth John | 77 Fighter/Interceptor Squadron |
| 39265 | 22.06.51 | 022136 | FLEMMING | James Hilary | 77 Fighter/Interceptor Squadron (shown as FLEMING for DFC (USA) award in LG39531) |
| | | 035498 | FOX | Raymond Francis | 77 Fighter/Interceptor Squadron |
| 39999 | 27.10.53 | 04424 | FRAYNE | Ross McRostie | 77 Fighter/Interceptor Squadron (shown as PRAYNE for MID award in LG39806) |
| 39265 | 22.06.51 | 023456 | FROST | Allen Kennedy | 77 Fighter/Interceptor Squadron |
| 39999 | 27.10.53 | 022133 | GARROWAY | William Maxwell | 77 Fighter/Interceptor Squadron |
| 39999 | 27.10.53 | 011402 | GOGERLEY | Bruce | 77 Fighter/Interceptor Squadron (shown as GOGERLY for DFC (USA) award in Australian listings and for DFC award in LG39661) |
| 39999 | 27.10.53 | 021993 | HAMILTON-FOSTER | Philip Vincent | 77 Fighter/Interceptor Squadron |
| | | 032533 | HOWARD | Hugh Bryan | 77 Fighter/Interceptor Squadron |
| | | 032378 | JANSON | Kenneth Robert | 77 Fighter/Interceptor Squadron |
| 39999 | 27.10.53 | 011417 | PHILP, DFC | Alan | 77 Fighter/Interceptor Squadron |
| 39265 | 22.06.51 | 05879 | READING | Leslie | 77 Fighter/Interceptor Squadron |
| 39265 | 22.06.51 | 033201 | THORNTON, DFM, AFM | Geoffrey | 77 Fighter/Interceptor Squadron |
| 39265 | 22.06.51 | 033260 | TREBILCO | Raymond Edward | 77 Fighter/Interceptor Squadron |
| 39265 | 22.06.51 | 022058 | WITTMAN | Richard William | 77 Fighter/Interceptor Squadron |
| PILOT OFFICERS | | | | | |
| 39265 | 22.06.51 | 022222 | COTTEE | Milton James | 77 Fighter/Interceptor Squadron |
| 39999 | 27.10.53 | 033831 | COWPER | Lionel Henry Cadogan | 77 Fighter/Interceptor Squadron |
| 39999 | 27.10.53 | 033821 | CRANSTON | Ian Rew | 77 Fighter/Interceptor Squadron (also listed as Sgt A33821 in Australian listings) |
| 39265 | 22.06.51 | 021261 | ELLIS | Donald Campbell | 77 Fighter/Interceptor Squadron (listed: missing. Also shown as Sgt A21261 in Australian listings) |
| 39265 | 22.06.51 | 04218 | KLAFFER | Lyall Robert | 77 Fighter/Interceptor Squadron |
| 39999 | 27.10.53 | 022790 | MURRAY, DFC, DFM | Kenneth James | 77 Fighter/Interceptor Squadron |
| 39999 | 27.10.53 | 022193 | MYERS, DFM | Edward John | 77 Fighter/Interceptor Squadron (shown as Sgt A22193 in Australian listings) |

| London Gazette issue No. | London Gazette date | Personal Service number | Name | Christian names | Remarks |
|---|---|---|---|---|---|
| AIR MEDAL (continued) | | | | | |
| PILOT OFFICERS (continued) | | | | | |
| 39999 | 27.10.53 | 033630 | OBORN, DFM | Ralph Victor | 77 Fighter/Interceptor Squadron |
| 39999 | 27.10.53 | 033838 | PARKER, DFM | John Norris | 77 Fighter/Interceptor Squadron (shown as Sgt A33838 in Australian listings) |
| 39999 | 27.10.53 | 032536 | ROBERTSON | Donald Neil | 77 Fighter/Interceptor Squadron (shown as Sgt A32536 in Australian listings) |
| 39999 | 27.10.53 | 022422 | ROBINSON | Richard George | 77 Fighter/Interceptor Squadron (shown as Sgt A22422 in Australian listings) |
| 39999 | 27.10.53 | 032542 | SIMMONDS | William Henry | 77 Fighter/Interceptor Squadron |
| 39999 | 27.10.53 | 033843 | SMITH | Kenneth Dudley | 77 Fighter/Interceptor Squadron (shown as Sgt A33843 in Australian listings) |
| 39265 | 22.06.51 | 0647 | STEPHENS | Geoffrey Ingram | 77 Fighter/Interceptor Squadron (shown as Sgt A647 in Australian listings) |
| 39999 | 27.10.53 | 033635 | STRAWBRIDGE, DFM | Robert Andrew | 77 Fighter/Interceptor Squadron (shown as Sgt A33635 in Australian listings) |
| 39999 | 27.10.53 | 032537 | SURMAN | John Leonard | 77 Fighter/Interceptor Squadron |
| 39999 | 27.10.53 | 032427 | THOMSON | Bruce Lachlan | 77 Fighter/Interceptor Squadron (listed: missing. Also shown as Sgt A32427 in Australian listings) |
| 39999 | 27.10.53 | 021121 | TOWNER, DFM | Kenneth Graham | 77 Fighter/Interceptor Squadron (shown as Sgt A21121 in Australian listings) |
| WARRANT OFFICERS | | | | | |
| 39265 | 22.06.51 | A22086 | BRACKENREG | Roydon Owen Leonard | 77 Fighter/Interceptor Squadron |
| 39265 | 22.06.51 | A528 | DOUGLAS | Eric Albert | 77 Fighter/Interceptor Squadron |
| 39265 | 22.06.51 | A33207 | FAIRWEATHER | Robert Forrester | 77 Fighter/Interceptor Squadron |
| 39999 | 27.10.53 | A11424 | HILL | Keith Gordon | 77 Fighter/Interceptor Squadron |
| 39265 | 22.06.51 | A33301 | HOWE | Charles Robert Albert | 77 Fighter/Interceptor Squadron |
| 39265 | 22.06.51 | A11389 | HUNT | Ronald Charles Arthur | 77 Fighter/Interceptor Squadron |
| 39265 | 22.06.51 | A11412 | MICHELSON | William Stapylton | 77 Fighter/Interceptor Squadron |
| 39265 | 22.06.51 | A22103 | NICHOLLS, DFM | Brian Frederick Stanley | 77 Fighter/Interceptor Squadron |
| | | A33275 | RAMSAY | D.A.K. | 77 Fighter/Interceptor Squadron |
| 39265 | 22.06.51 | A22108 | RIVERS | Wallace Bolton | 77 Fighter/Interceptor Squadron (shown as Flying Officer 02218 in Australian listings) |

| London Gazette issue No. | London Gazette date | Personal Service number | Name | Christian names | Remarks |
|---|---|---|---|---|---|
| AIR MEDAL (continued) | | | | | |
| WARRANT OFFICERS (continued) | | | | | |
| 39265 | 22.06.51 | A22168 | TURNER | Avenal Richard | 77 Fighter/Interceptor Squadron |
| 39265 | 22.06.51 | A33213 | WILLIAMSON | Stanley William | 77 Fighter/Interceptor Squadron |
| PILOT III (SERGEANT) | | | | | |
| 39265 | 22.06.51 | A33272 | HARROP | William Percy | 77 Fighter/Interceptor Squadron (listed: since deceased) |
| FLIGHT-SERGEANTS | | | | | |
| 39999 | 27.10.53 | A22221 | ARMIT | Ernest Donald | 77 Fighter/Interceptor Squadron (listed: missing) |
| 39999 | 27.10.53 | A22140 | AVERY, DFM | Allan James | 77 Fighter/Interceptor Squadron |
| 39999 | 27.10.53 | A22218 | COLLINS | Frederick Thomas | 77 Fighter/Interceptor Squadron |
| 39999 | 27.10.53 | A51083 | MIDDLEMISS | William | 77 Fighter/Interceptor Squadron |
| SERGEANTS | | | | | |
| 39999 | 27.10.53 | A34233 | ARMSTRONG, DFM | Tony Charles | 77 Fighter/Interceptor Squadron |
| 39265 | 22.06.51 | A32245 | BESSELL | Henry William | 77 Fighter/Interceptor Squadron |
| 39999 | 27.10.53 | A33621 | BLACKWELL | Frank Robert | 77 Fighter/Interceptor Squadron |
| 39999 | 27.10.53 | A5895 | COLEBROOK, DFM | Maxwell Edwin | 77 Fighter/Interceptor Squadron |
| 39999 | 27.10.53 | A33624 | DRUMMOND | Vance | 77 Fighter/Interceptor Squadron (listed: missing) |
| 39999 | 27.10.53 | A22613 | EVANS | John Alexander | 77 Fighter/Interceptor Squadron |
| 39265 | 22.06.51 | A4443 | FOSTER | Kevin Henry | 77 Fighter/Interceptor Squadron |
| 39999 | 27.10.53 | A33625 | GILLAN | Bruce Thompson | 77 Fighter/Interceptor Squadron (listed: missing) |
| 39265 | 22.06.51 | A33280 | HUNT | Herbert Ronald | 77 Fighter/Interceptor Squadron |
| 39999 | 27.10.53 | A33628 | KITCHENSIDE | James Charles | 77 Fighter/Interceptor Squadron |
| 39999 | 27.10.53 | A33830 | KING, DFM | Colin George | 77 Fighter/Interceptor Squadron |
| 39999 | 27.10.53 | A34254 | LUSHEY, DFM | Geoffrey Warren | 77 Fighter/Interceptor Squadron |
| 39265 | 22.06.51 | A33283 | MEGGS | Keith Raymond | 77 Fighter/Interceptor Squadron |
| 39265 | 22.06.51 | A22223 | MITCHELL | Ronald Daniel | 77 Fighter/Interceptor Squadron |
| 39999 | 27.10.53 | A33837 | OUTHWAITE | Maxwell Allan | 77 Fighter/Interceptor Squadron |
| 39999 | 27.10.53 | A33296 | ROBSON | Roy | 77 Fighter/Interceptor Squadron (listed: since deceased) |
| 39265 | 22.06.51 | A33217 | ROYAL | Kenneth Edward | 77 Fighter/Interceptor Squadron (listed: since deceased) |
| 39265 | 22.06.51 | A33287 | SLY | Cecil | 77 Fighter/Interceptor Squadron |
| 39265 | 22.06.51 | A22110 | STONEY | Alfred Thomas | 77 Fighter/Interceptor Squadron |
| 39999 | 27.10.53 | A11439 | ZUPP | Phillip | 77 Fighter/Interceptor Squadron |

NEW ZEALAND

| London Gazette issue No. | London Gazette date | Personal Service number | Name | Christian names | Remarks |
|---|---|---|---|---|---|
| ROYAL NEW ZEALAND NAVY | | | | | |
| MBE: MEMBER OF THE ORDER OF THE BRITISH EMPIRE | | | | | |
| LIEUTENANT (E) | | | | | |
| 40011 | 06.11.53 | not listed | SIMMONDS | Henry Rogerson | HMNZS Hawea |
| 2ND BAR TO DSC: DISTINGUISHED SERVICE CROSS | | | | | |
| LIEUTENANT COMMANDER | | | | | |
| | | not listed | GRAHAME, DSC | G.O. | HMNZS Rotoiti |
| BAR TO DSC: DISTINGUISHED SERVICE CROSS | | | | | |
| COMMANDER | | | | | |
| 39870 | 01.06.53 | not listed | DAVIS-GOFF, DSC | George Raymond | HMNZS Hawea |
| DSC: DISTINGUISHED SERVICE CROSS | | | | | |
| COMMANDER | | | | | |
| 39528 | 23.05.52 | not listed | TURNER | Brian Edmund | HMNZS Rotoiti |
| LIEUTENANT COMMANDERS | | | | | |
| 40109 | 19.02.54 | not listed | CARR | Lawrence George | HMNZS Kaniere |
| 39660 | 03.10.52 | not listed | JOHNSTON | Francis Nigel Featherstone | HMNZS Hawea |
| SENIOR COMMISSIONED ELECTRICAL OFFICER (L) | | | | | |
| 39528 | 23.05.52 | not listed | GRANT, BEM | Michael | HMNZS Rotoiti |
| DSM: DISTINGUISHED SERVICE MEDAL | | | | | |
| ABLE SEAMEN | | | | | |
| 39547 | 23.05.52 | NZ11418 | BUTTON | Edward James | HMNZS Rotoiti |
| 39547 | 23.05.52 | NZ11321 | SCOLES | Norman Joseph | HMNZS Rotoiti |
| MENTION IN DESPATCHES (FOR GALLANT AND DISTINGUISHED SERVICES - POSTHUMOUS) | | | | | |
| BRONZE OAK LEAF EMBLEM | | | | | |
| ABLE SEAMAN | | | | | |
| 39547 | 23.05.52 | NZ13155 | MARCHIONI | Robert Edward | HMNZS Rotoiti |
| MENTION IN DESPATCHES (FOR GALLANT AND DISTINGUISHED SERVICES) | | | | | |
| BRONZE OAK LEAF EMBLEM | | | | | |
| COMMANDER | | | | | |
| | | not listed | HERRICK, DSC | L.E. | HMNZS Pukaki |
| LIEUTENANT COMMANDER | | | | | |
| | | not listed | HOARE | P.J. | HMNZS Tutira |

| London Gazette issue No. | London Gazette date | Personal Service number | Name | Christian names | Remarks |
|---|---|---|---|---|---|
| MENTION IN DESPATCHES (FOR GALLANT AND DISTINGUISHED SERVICES) (continued) | | | | | |
| BRONZE OAK LEAF EMBLEM | | | | | |
| SURGEON LIEUTENANT | | | | | |
| 39725 | 23.12.52 | not listed | FORSYTH, MB, Ch.B | James Ian McLeish | HMNZS Taupo |
| LIEUTENANT | | | | | |
| 39854 | 19.05.53 | not listed | DAVIES | David Hilary | HMNZS Rotoiti |
| SENIOR COMMISSIONED MECHANICIAN | | | | | |
| 39660 | 03.10.52 | not listed | MITCHELL | George | HMNZS Hawea |
| CHIEF PETTY OFFICER STOKER MECHANIC | | | | | |
| 40011 | 06.11.53 | NZ12804 | CUNNINGHAM | Kenneth Frank | HMNZS Hawea |
| 39725 | 23.12.52 | NZ12327 | LINES | Prentice Alfred | HMNZS Taupo |
| CHIEF PETTY OFFICER TELEGRAPHIST | | | | | |
| 40011 | 06.11.53 | NZ12043 | BICKLEY | Douglas Gordon | HMNZS Hawea |
| CHIEF PETTY OFFICERS | | | | | |
| 39660 | 03.12.52 | NZ13108 | FLETCHER | Harold George Sydney | HMNZS Hawea |
| 39725 | 23.12.52 | NZ1281 | TAYLOR | Amos Richard | HMNZS Taupo |
| ORDNANCE ARTIFICER (2) | | | | | |
| 39870 | 01.06.53 | NZ12548 | GRAY | John Findlay | HMNZS Rotoiti (also listed as GREY) |
| ELECTRICAL ARTIFICER (3) | | | | | |
| 39336 | 19.09.51 | NZ12249 | MOFFAT | Robert Thomas Noble | HMNZS Pukaki |
| PETTY OFFICER TELEGRAPHIST | | | | | |
| 39547 | 23.05.52 | NZ2102 | BOYLAND | James Albert | HMNZS Rotoiti |
| PETTY OFFICER COOK (S) | | | | | |
| 39547 | 23.05.52 | NZ13090 | JENKINS | Caleb Charles | HMNZS Tutira |
| LEADING ELECTRICIAN'S MATE | | | | | |
| 39236 | 25.05.51 | NZ11365 | SOWDEN | Walter Neil | ∗ Ship not listed |
| 39854 | 19.05.53 | NZ11703 | WENSLEY | Louis Alfred | HMNZS Kaniere (also listed as WENSKY) |
| LEADING SEAMAN | | | | | |
| 40109 | 19.02.54 | NZ12371 | CLEARY | David Joseph | HMNZS Kaniere |

ROYAL NEW ZEALAND ARMY

| London Gazette issue No. | London Gazette date | Personal Service number | Name | Christian names | Remarks |
|---|---|---|---|---|---|

HEADQUARTERS/STAFF

CB: COMPANION OF THE ORDER OF THE BATH

BRIGADIER

| London Gazette issue No. | London Gazette date | Personal Service number | Name | Christian names | Remarks |
|---|---|---|---|---|---|
| 39756 | 20.01.53 | 30004 | PARK, CBE | Ronald Stewart | Staff Corps |

| London Gazette issue No. | London Gazette date | Personal Service number | Name | Christian names | Remarks |
|---|---|---|---|---|---|
| ROYAL NEW ZEALAND ARTILLERY | | | | | |
| DSO: DISTINGUISHED SERVICE ORDER | | | | | |
| LIEUTENANT-COLONELS | | | | | |
| 40161 | 30.04.54 | 30055 | BURNS. MBE | John | Officer Commanding 16th Field Regiment |
| 39442 | 18.01.52 | 200042 | MOODIE, ED | John William | Officer Commanding 16th Field Regiment |
| 39892 | 19.06.53 | 30056 | PATERSON | McKay | Officer Commanding 16th Field Regiment (also listed as Robertson McKay) (also listed as PETERSON) |
| CAPTAIN | | | | | |
| 39494 | 18.03.52 | 206382 | KING, MC | Peter Frank | 16th Field Regiment |
| MBE: MEMBER OF THE ORDER OF THE BRITISH EMPIRE | | | | | |
| MAJORS | | | | | |
| | | not listed | DIXON | W.G. | 16th Field Regiment |
| 39756 | 20.01.53 | 202250 | SOLOMON | George | 16th Field Regiment |
| CAPTAINS | | | | | |
| 39870 | 01.06.53 | 203512 | CHANNINGS | Andrew | 16th Field Regiment |
| | | 11120 | WEIR | D.D. | 16th Field Regiment |
| MC: MILITARY CROSS | | | | | |
| MAJORS | | | | | |
| 40012 | 10.11.53 | 206081 | DULEY | Vernon James | 16th Field Regiment |
| 40012 | 10.11.53 | 30167 | MANDERS | Ernest John | 16th Field Regiment |
| 39870 | 01.06.53 | not listed | SKILTON | Vincent George | 16th Field Regiment |
| 40161 | 30.04.54 | 30156 | SPENCE | James Roy | 16th Field Regiment |
| CAPTAINS | | | | | |
| | | 206382 | KING, DSO | Peter Frank | 16th Field Regiment (initials also listed as P.E.) |
| 39442 | 18.01.52 | 206085 | MASON | Ronald Edgar | 16th Field Regiment |
| 39456 | 05.02.52 | 206084 | MILLER | Norman Lawrence | 16th Field Regiment |
| 39582 | 24.06.52 | 206384 | RESTON | Robert Murray | 16th Field Regiment |
| 39207 | 20.04.51 | 206379 | ROXBURGHE | Arthur A. | 16th Field Regiment (also listed as ROXBURGH) |
| 39892 | 19.06.53 | 207885 | STANAWAY | Murray Clifford | 16th Field Regiment |
| LIEUTENANT | | | | | |
| 39597 | 15.07.52 | 203627 | HENDRY | Wilmot | 16th Field Regiment |

| London Gazette issue No. | London Gazette date | Personal Service number | Name | Christian names | Remarks |
|---|---|---|---|---|---|
| DCM: DISTINGUISHED CONDUCT MEDAL | | | | | |
| GUNNER | | | | | |
| 39494 | 18.03.52 | 206285 | RIXON | Derek Edwin | 16th Field Regiment |
| MM: MILITARY MEDAL | | | | | |
| SERGEANTS | | | | | |
| 39892 | 19.06.53 | 207014 | REDFEARN | Bruce Andrew | 16th Field Regiment |
| 39892 | 19.06.53 | 206204 | REID | James Reid | 16th Field Regiment (listed: killed in action 02.05.53) |
| BOMBARDIER | | | | | |
| 39756 | 20.01.53 | 207853 | GORDON | Louis John Maioni | 16th Field Regiment |
| LANCE-BOMBARDIERS | | | | | |
| 39456 | 05.02.52 | 207782 | BUCHANAN | Norman James | 16th Field Regiment |
| 39456 | 05.02.52 | 207731 | JONES | L. Raymond | 16th Field Regiment |
| 39273 | 29.06.51 | 203851 | McCUBBIN | Hector Keith | 16th Field Regiment (also listed as McGUBBIN) |
| GUNNER | | | | | |
| 39883 | 09.06.53 | 208057 | CLARKE | William Leslie | 16th Field Regiment |
| BEM: BRITISH EMPIRE MEDAL | | | | | |
| STAFF SERGEANT | | | | | |
| | | not listed | SMITH | R.M. | 16th Field Regiment |
| GUNNER | | | | | |
| NZG No. not listed | 10.01.52 | not listed | BLUETT | Max | 16th Field Regiment |
| MENTION IN DESPATCHES (FOR GALLANT AND DISTINGUISHED SERVICES - POSTHUMOUS) | | | | | |
| BRONZE OAK LEAK EMBLEM | | | | | |
| 2ND LIEUTENANT | | | | | |
| 39331 | 11.09.51 | 203624 | FIELDEN | D.S. | 16th Field Regiment (listed: killed in action 24.04.51) |
| MENTION IN DESPATCHES (FOR GALLANT AND DISTINGUISHED SERVICES) | | | | | |
| BRONZE OAK LEAF EMBLEM | | | | | |
| MAJORS | | | | | |
| 39892 | 19.06.53 | 30173 | HASSETT | R.D.P. | 16th Field Regiment |
| 39331 | 11.09.51 | 206375 | HUNT | E.W. | 16th Field Regiment |
| 40161 | 30.04.54 | 206084 | MILLER, MC | N.L. | 16th Field Regiment |
| 39484 | 04.03.52 | 202241 | NATHAN | H.G. | 16th Field Regiment |
| 39484 | 04.03.52 | 30099 | WEBB | R.J.H. | 16th Field Regiment (also listed: Officer Commanding Lt-Col.) |

| London Gazette issue No. | London Gazette date | Personal Service number | Name | Initials | Remarks |
|---|---|---|---|---|---|
| MENTION IN DESPATCHES (FOR GALLANT AND DISTINGUISHED SERVICES) (continued) | | | | | |
| BRONZE OAK LEAF EMBLEM | | | | | |
| CAPTAINS | | | | | |
| 39484 | 04.03.52 | 206380 | CHESSUM | R.S. | 16th Field Regiment |
| 39582 | 24.06.52 | 30222 | JOPLIN | P.W.F. | 16th Field Regiment |
| 39756 | 20.01.53 | 206088 | MOLOUGHNEY | C.J. | 16th Field Regiment |
| 39448 | 04.03.52 | 30231 | PORTER | R.K.G. | 16th Field Regiment |
| 39756 | 20.01.53 | 31037 | VINE | T.A.N. | 16th Field Regiment |
| LIEUTENANTS | | | | | |
| 39892 | 19.06.53 | 203713 | HICKEY | W.J. | 16th Field Regiment |
| 39484 | 04.03.52 | 207685 | SCOTT | D.J. | 16th Field Regiment |
| 2ND LIEUTENANTS | | | | | |
| 39582 | 24.06.52 | 203709 | GRIFFITHS | H.K. | 16th Field Regiment |
| 39582 | 24.06.52 | 203627 | HENDRY, MC | W. | 16th Field Regiment |
| WARRANT OFFICER CLASS I | | | | | |
| 39756 | 20.01.53 | 203312 | DICKINSON | I.J. | 16th Field Regiment |
| BOMBARDIERS | | | | | |
| 39756 | 20.01.53 | 296232 | BEANGE | J.H. | 16th Field Regiment |
| 39282 | 19.06.53 | 207940 | BELL | W.J. | 16th Field Regiment |
| 39484 | 04.03.52 | 203926 | CARE | C. | 16th Field Regiment |
| 39892 | 19.06.53 | 206293 | THOMPSON | G.H. | 16th Field Regiment |
| LANCE-BOMBARDIERS | | | | | |
| 40012 | 10.11.53 | 203629 | CONWAY | K.L. | 16th Field Regiment |
| 39582 | 24.06.52 | 207921 | POLLARD | G.W. | 16th Field Regiment |
| GUNNERS | | | | | |
| 40161 | 30.04.54 | 206616 | BOLTON | A. | 16th Field Regiment (initials also shown as G.A.) |
| 40161 | 30.04.54 | 206907 | MacLEOD | T. | 16th Field Regiment |
| 39871 | 26.05.53 | not listed | REID | J.F. | 16th Field Regiment |
| 39871 | 26.05.53 | not listed | RYAN | C.E. | 16th Field Regiment |
| 39484 | 04.03.52 | 203813 | SEMMENS | M.K. | 16th Field Regiment |
| 40012 | 10.11.53 | 204277 | TAIMANA | H.P. | 16th Field Regiment (initials also shown as H.R.) |
| 39582 | 24.06.52 | 206295 | WILSON | G.G. | 16th Field Regiment |

AWARD CONFERRED BY THE PRESIDENT OF THE UNITED STATES OF AMERICA

| London Gazette issue No. | London Gazette date | Personal Service number | Name | Initials | Remarks |
|---|---|---|---|---|---|
| AIR MEDAL | | | | | |
| LIEUTENANT | | | | | |
| | | not listed | WATSON | W. | 16th Field Regiment |

| London Gazette issue No. | London Gazette date | Personal Service number | Name | Christian names | Remarks |
|---|---|---|---|---|---|

ROYAL NEW ZEALAND CORPS OF ENGINEERS

MC: MILITARY CROSS

2ND LIEUTENANT

| London Gazette issue No. | London Gazette date | Personal Service number | Name | Christian names | Remarks |
|---|---|---|---|---|---|
| 39892 | 19.06.53 | 207544 | BUTCHER | George William | ← Unit not listed |

BEM : BRITISH EMPIRE MEDAL

SAPPER

| London Gazette issue No. | London Gazette date | Personal Service number | Name | Christian names | Remarks |
|---|---|---|---|---|---|
| 40012 | 10.11.53 | 204188 | PITMAN | Edward Walter | ← |

MENTION IN DESPATCHES (FOR GALLANT AND DISTINGUISHED SERVICES)
BRONZE OAK LEAF EMBLEM

2ND LIEUTENANTS

| London Gazette issue No. | London Gazette date | Personal Service number | Name | Christian names | Remarks |
|---|---|---|---|---|---|
| 39892 | 19.06.53 | 203931 | HALL | K.O. | New Zealand Base Headquarters (initials also listed as K.G.) |
| 39494 | 04.03.52 | 206023 | VELVIN | M.N. | ← |

← Unit not listed

| London Gazette issue No. | London Gazette date | Personal Service number | Name | Christian names | Remarks |
|---|---|---|---|---|---|
| **ROYAL NEW ZEALAND CORPS OF SIGNALS** | | | | | |
| **MBE: MEMBER OF THE ORDER OF THE BRITISH EMPIRE** | | | | | |
| **LIEUTENANT** | | | | | |
| 40012 | 10.11.53 | 206943 | HILL | William Murray | att 16th Field Regiment RNZA |
| **2ND LIEUTENANT** | | | | | |
| 30756 | 20.01.53 | 207894 | BURROWS | Richard Marchant | att 16th Field regiment RNZA |
| **MENTION IN DESPATCHES (FOR GALLANT AND DISTINGUISHED SERVICES)** | | | | | |
| **BRONZE OAK LEAF EMBLEM** | | | | | |
| **CAPTAIN** | | | | | |
| 39892 | 19.06.53 | 31353 | PENLINGTON | P.B. | ← Unit not listed |
| **LIEUTENANT** | | | | | |
| 40012 | 10.11.53 | 202562 | GRIGG | P.J. | ← |
| **SERGEANTS** | | | | | |
| 40012 | 10.11.53 | 203994 | CAVAYE | K.W. | ← (also listed as a L/CPL) |
| 39892 | 19.06.53 | 206152 | DUNFORD | I.E. | Headquarters K Force |
| 39756 | 20.01.53 | 206585 | WALLACE | R. | ← |
| **CORPORALS** | | | | | |
| 40012 | 10.11.53 | 204203 | HISHON | J. | ← |
| 39892 | 19.06.53 | 208145 | MacDONALD | R.G. | ← (also listed as MacDONNELL) |
| 39756 | 20.01.53 | 208251 | STRUTHERS | R.A. | ← |
| **LANCE-CORPORAL** | | | | | |
| 39892 | 19.06.53 | 204015 | McINTYRE | B. | ← |

← Unit not listed

| London Gazette issue No. | London Gazette date | Personal Service number | Name | Christian names | Remarks |
|---|---|---|---|---|---|
| ROYAL NEW ZEALAND REGIMENT | | | | | |
| MBE: MEMBERS OF THE ORDER OF THE BRITISH EMPIRE | | | | | |
| MAJOR | | | | | |
| | | not listed | MUSGRAVE | L.R. | att 16th Field Regiment RNZA |
| CAPTAIN | | | | | |
| | | not listed | LLEWELLYN | S.P. | att 16th Field Regiment RNZA |
| MENTION IN DESPATCHES (FOR GALLANT AND DISTINGUISHED SERVICES) BRONZE OAK LEAF EMBLEM | | | | | |
| CAPTAIN | | | | | |
| 39871 | 26.05.53 | not listed | POANANGA | B.M. | att 3rd Bn Royal Australian Regiment |
| WARRANT OFFICER CLASS I | | | | | |
| 39756 | 20.01.53 | 33957 | GUMMER | E.C. | Headquarters K Force |
| SERGEANT | | | | | |
| 39871 | 26.05.53 | not listed | SPIERS | J.M. | att 3rd Bn Royal Australian Regiment |

| London Gazette issue No. | London Gazette date | Personal Service number | Name | Christian names | Remarks |
|---|---|---|---|---|---|
| ROYAL NEW ZEALAND ARMY SERVICE CORPS | | | | | |
| MBE: MEMBER OF THE ORDER OF THE BRITISH EMPIRE | | | | | |
| MAJOR | | | | | |
| 39892 | 19.06.53 | 202561 | COOPER | Alan William | 10 Company |
| CAPTAINS | | | | | |
| 40161 | 30.04.54 | 203800 | FOTHERINGHAME | Austin Sinclair | 10 Company (also listed FOTHERINGHOME) |
| | | not listed | MALLASCH | T.J.W. | attached Base Pay |
| BEM: BRITISH EMPIRE MEDAL | | | | | |
| STAFF SERGEANT | | | | | |
| 40012 | 10.11.53 | 208304 | O'BRIEN | Ronald Ferguson | 10 Company |
| MENTION IN DESPATCHES (FOR GALLANT AND DISTINGUISHED SERVICES) BRONZE OAK LEAF EMBLEM | | | | | |
| MAJOR | | | | | |
| 40161 | 30.04.54 | 30093 | MABBETT | J.M. | ← Unit not listed |
| SERGEANTS | | | | | |
| 40012 | 10.11.53 | 206738 | COULTER | D. McB. | ← |
| 40012 | 10.11.53 | 206501 | JORDAN | T. | ← (initial also listed as R.) |
| LANCE-CORPORAL | | | | | |
| 39756 | 20.01.53 | 203825 | HAWKES | I.L. | ← |

← Unit not listed

| London Gazette issue No. | London Gazette date | Personal Service number | Name | Christian names | Remarks |
|---|---|---|---|---|---|
| ROYAL NEW ZEALAND ARMY MEDICAL CORPS | | | | | |
| MENTION IN DESPATCHES (FOR GALLANT AND DISTINGUISHED SERVICES) BRONZE OAK LEAF EMBLEM | | | | | |
| STAFF SERGEANT | | | | | |
| 40012 | 10.11.53 | 203689 | LONG | S.V. | att 16th Field Regiment RNZA |
| SERGEANT | | | | | |
| 39756 | 20.01.53 | 207711 | BENYON | R.T. | att 16th Field Regiment RNZA |

| London Gazette issue No. | London Gazette date | Personal Service number | Name | Christian names | Remarks |
|---|---|---|---|---|---|
| ROYAL NEW ZEALAND ARMY ORDNANCE CORPS | | | | | |
| MBE: MEMBER OF THE ORDER OF THE BRITISH EMPIRE | | | | | |
| MAJOR | | | | | |
| 39892 | 19.06.53 | 30431 | ATKINSON | Geoffrey John Hayes | att 16th Field Regiment RNZA and HQ 1st Commonwealth Division |
| BEM: BRITISH EMPIRE MEDAL | | | | | |
| SERGEANT | | | | | |
| 40012 | 10.11.53 | 206870 | DON | James Russell | Central Ammunition Depot, Korea (also listed as Private) |

| London Gazette issue No. | London Gazette date | Personal Service number | Name | Christian names | Remarks |
|---|---|---|---|---|---|
| ROYAL NEW ZEALAND CORPS OF ELECTRICAL AND MECHANICAL ENGINEERS | | | | | |
| MBE: MEMBER OF THE ORDER OF THE BRITISH EMPIRE | | | | | |
| CAPTAIN | | | | | |
| 39756 | 20.01.53 | 206377 | WILSON | Jack Mahoney | att 16th Field Regiment RNZA |

| London Gazette issue No. | London Gazette date | Personal Service number | Name | Christian names | Remarks |
|---|---|---|---|---|---|

ROYAL NEW ZEALAND ARMY DENTAL CORPS

MBE: MEMBER OF THE ORDER OF THE BRITISH EMPIRE

CAPTAIN

| London Gazette issue No. | London Gazette date | Personal Service number | Name | Christian names | Remarks |
|---|---|---|---|---|---|
| 39892 | 19.06.53 | 202452 | CULL | Alan Hunter | att 16th Field Regiment RNZA |

INDIA

| India Gazette issue No. | India Gazette date | Personal Service number | Name | Other names | Remarks |
|---|---|---|---|---|---|
| INDIAN ARMY | | | | | |
| AWARD: MAHA-VIR CHAKRA | | | | | |
| LIEUTENANT-COLONEL | | | | | |
| 38-PUC | 25.08.51 | MR-252 | RANGARAJ | Arcot Govindaraj | Indian Army Medical Corps Officer Commanding 60th Para Field Ambulance |
| MAJOR | | | | | |
| 5-PRES | 19.01.52 | MR-281 | BANERJEE | Narenda Baran | Indian Army Medical Corps |
| AWARD: BAR TO VIR CHAKRA | | | | | |
| MAJOR | | | | | |
| 6-PRES | 19.01.52 | MR-286 | RANGASWAMI | Venkata Pathi | Indian Army Medical Corps |
| AWARD: VIR CHAKRA | | | | | |
| CAPTAINS | | | | | |
| | | MR-381 | BANERJEE | Ashok | Indian Army Medical Corps |
| 6-PRES | 19.01.52 | MR-497 | DAS | Narenda Chandra | Indian Army Medical Corps 60th Para Field Ambulance |
| NAIK | | | | | |
| 21-PRES | 25.07.53 | 29485 | Nagsen SINGH | | Indian Army Medical Corps 60th Para Field Ambulance (also listed as Dubedar/ORA |
| 6-PRES | 19.01.52 | 43424 | Ratan SINGH | | Indian Army Medical Corps |
| 21-PRES | 25.07.53 | 6574258 | Umrao SINGH | | Indian Army Service Corps (MT) |
| LANCE-NAIK | | | | | |
| | | 6564273 | Budh SINGH | | Indian Army Service Corps |

| India Gazette issue No. | India Gazette date | Personal Service number | Name | Other names | Remarks |
|---|---|---|---|---|---|
| MENTION IN DESPATCHES (FOR GALLANT AND DISTINGUISHED SERVICES) | | | | | |
| BRONZE OAK LEAF EMBLEM | | | | | |
| LIEUTENANT-COLONEL | | | | | |
| | | MR-252 | RANGARAJ | Arcot Govindaraj | Indian Army Medical Corps Officer Commanding 60th Para Field Ambulance |
| MAJOR | | | | | |
| | | MR-286 | RANGASWAMI | Venkata Pathi | Indian Army Medical Corps Twice Mentioned in Despatches |
| CAPTAINS | | | | | |
| | | MR-381 | BANERJEE | Ashok | Indian Army Medical Corps |
| 7-PRES | 19.01.52 | MR-397 | BASU | Amulya Kumar | Indian Army Medical Corps |
| 16-PRES | 25.06.53 | M-21140 | DAS GUPTA | Parimal Chandra | Indian Army Medical Corps |
| 39-PRES | 11.06.52 | MS-5203 | KRISHNAN | Subbu | Indian Army Medical Corps |
| 39-PRES | 11.06.52 | SS-14594 | Joginder SINGH | | Listed PUNJAB |
| LIEUTENANT | | | | | |
| 39-PRES | 11.06.52 | IEC-5203 | RAMANA | Y.V. | Indian Army Service Corps |
| SUBEDAR | | | | | |
| 39-PRES | 11.06.52 | 31543 | CHANDOLA | Laxshmi Datt | Indian Army Medical Corps |
| NURSING NAIK | | | | | |
| 39-PRES | 11.06.52 | 29485 | Nagsen SINGH | | Indian Army Medical Corps |
| NAIK-AMB | | | | | |
| 16-PRES | 25.06.53 | 94328 | KAPTHINGA | | Indian Army Medical Corps |
| 15-PRES | 08.03.54 | 15818 | RAMASWAMY | S. | IndianArmy Medical Corps |
| LANCE-NAIK | | | | | |
| 7-PRES | 19.01.52 | 66242 | ASREY | Ram | Indian Army Medical Corps |
| 7-PRES | 19.01.52 | 34676 | SURAMAN | Ram | Indian Army Medical Corps |
| 14-PRES | 07.05.53 | 6754744 | NARAYANAN | | Indian Army Medical Corps |
| SEPOY | | | | | |
| 7-PRES | 19.01.52 | 6539267 | DATT | Hari | Indian Army Service Corps |

| India Gazette issue No. | India Gazette date | Personal Service number | Name | Other Names | Remarks |
|---|---|---|---|---|---|

MENTION IN DESPATCHES (FOR GALLANT AND DISTINGUISHED SERVICES)(Continued)
BRONZE OAK LEAF EMBLEM

| India Gazette issue No. | India Gazette date | Personal Service number | Name | Other Names | Remarks |
|---|---|---|---|---|---|
| SEPOY/DVR | | | | | |
| 16-PRES | 25.06.53 | 6574251 | VALLIATHAN | P. | Indian Army Medical Corps |
| 15-PRES | 08.03.54 | 6548395 | RAM | Raja | Indian Army Service Corps |
| SEPOY/NO | | | | | |
| 15-PRES | 08.03.54 | 6745738 | CHERIAN | M. | Indian Army Medical Corps |
| SEPOY/AO | | | | | |
| 16-PRES | 25.06.53 | 6728190 | SACHIDANAND | | Indian Army Medical Corps |
| JAMADAR | | | | | |
| 39-PRES | 11.06.52 | 50587-10 | TEWARI | Purna Nand | Indian Army Medical Corps |
| JAMADAR/MT | | | | | |
| 16-PRES | 25.06.53 | 39570 | SHARMA | Narayan Datt | Indian Army Medical Corps |
| 14-PRES | 07.05.53 | 42266-10 | SHETTY | Gopal | Indian Army Service Corps |
| 14-PRES | 07.05.53 | 49843-10 | Mohan Lal | | Indian Army Medical Corps |
| AMB HAV | | | | | |
| 39-PRES | 11.06.52 | 92937 | Nain SINGH | | Indian Army Medical Corps |
| HAV | | | | | |
| 15-PRES | 08.03.54 | 87215 | THARAKHAN | M.S. | Indian Army Medical Corps |

# SOUTH AFRICA

| London Gazette issue No. | London Gazette date | Personal Service number | Name | Initials | Remarks |
|---|---|---|---|---|---|

SOUTH AFRICAN ARMY

SOUTH AFRICAN ARMOURED CORPS

MENTION IN DESPATCHES (FOR GALLANT AND DISTINGUISHED SERVICES) BRONZE OAK LEAF EMBLEM

MAJORS

| | | | | | |
|---|---|---|---|---|---|
| 40262 | 24.08.54 | not listed | HAMMON | D.J. | ← Unit not listed (also shown as HAMMAN) |
| 40262 | 24.08.54 | not listed | SCHUMAN | C.H. | ← Unit not listed |

| London Gazette issue No. | London Gazette date | Personal Service number | Name | Christian names | Remarks |
|---|---|---|---|---|---|

SOUTH AFRICAN ARTILLERY

MBE: MEMBER OF THE ORDER OF THE BRITISH EMPIRE

CAPTAIN

| London Gazette issue No. | London Gazette date | Personal Service number | Name | Christian names | Remarks |
|---|---|---|---|---|---|
| 40262 | 24.08.54 | not listed | GIBBS | Winton Mortimer | att HQ 1st Commonwealth Division |

# NON-OPERATIONAL SERVICES IN JAPAN IN CONNECTION WITH UNITED NATIONS OPERATIONS DURING KOREAN WAR

## AWARDS

| London Gazette issue No. | London Gazette date | Personal Service number | Name | Christian names | Remarks |
|---|---|---|---|---|---|
| ROYAL CANADIAN ARMY | | | | | |
| HEADQUARTERS/STAFF | | | | | |
| CBE: COMMANDER OF THE ORDER OF THE BRITISH EMPIRE | | | | | |
| BRIGADIER | | | | | |
| 39560 | 30.05.52 | ZD637 | FLEURY, MBE, ED | Frank James | Staff Corps |

| London Gazette issue No. | London Gazette date | Personal Service number | Name | Christian names | Remarks |
|---|---|---|---|---|---|

ROYAL CANADIAN ARMOURED CORPS

OBE: OFFICER OF THE ORDER OF THE BRITISH EMPIRE

LIEUTENANT-COLONEL

| London Gazette issue No. | London Gazette date | Personal Service number | Name | Christian names | Remarks |
|---|---|---|---|---|---|
| 39560 | 30.01.52 | ZD1026 | SARE | Paul Francis Lionel | BCOF |

| London Gazette issue No. | London Gazette date | Personal Service number | Name | Christian names | Remarks |
|---|---|---|---|---|---|

ROYAL CANADIAN ARMY INFANTRY CORPS

BEM: BRITISH EMPIRE MEDAL

SERGEANT

| London Gazette issue No. | London Gazette date | Personal Service number | Name | Christian names | Remarks |
|---|---|---|---|---|---|
| 39738 | 01.01.53 | SK62292 | TUTTE | Kenneth Gordon | ← Unit not listed |

| London Gazette issue No. | London Gazette date | Personal Service number | Name | Christian names | Remarks |
|---|---|---|---|---|---|
| ROYAL CANADIAN ARMY SERVICE CORPS | | | | | |
| BEM: BRITISH EMPIRE MEDAL | | | | | |
| STAFF-SERGEANT | | | | | |
| 39560 | 30.05.52 | SG9020 | CLOUSTON | Byron Clifford | ← Unit not listed |

| Canadian/ London Gazette issue No. | Canadian/ London Gazette date | Personal Service number | Name | Initials | Remarks |
|---|---|---|---|---|---|
| ROYAL CANADIAN AIR FORCE | | | | | |
| KING'S COMMENDATION FOR BRAVE CONDUCT BRONZE OAK LEAF EMBLEM | | | | | |
| CORPORAL | | | | | |
| 51/259 | | 24985 | TALSON | E.H. | ← Unit not listed |
| KING'S COMMENDATION FOR VALUABLE SERVICES IN THE AIR BRONZE OAK LEAF EMBLEM | | | | | |
| SQUADRON LEADER | | | | | |
| 51/395 | | 20215 | EVANS | T.J. | ← |
| QUEEN'S COMMENDATION FOR VALUABLE SERVICES IN THE AIR BRONZE OAK LEAF EMBLEM | | | | | |
| FLIGHT LIEUTENANTS | | | | | |
| 52/406 | | 23526 | ENDERSBEE | C.E. | ← |
| 52/406 | | 6552 | FICKLESTEIN | A. | ← |
| 53/662 | | 18115 | JANZEN | R.H. | ← |
| not listed | | 17600 | MILLER | J.B. | ← |
| 52/406 | | 17706 | RATCLIFFE | R.E.D. | ← |
| 52/406 | | 20405 | SMITH | W. | ← |
| SERGEANTS | | | | | |
| 52/406 | | 22554 | BOWMAN | F.S.M. | ← |
| 53/360 | | 22223 | McKNIGHT | W.S. | ← |
| CORPORAL | | | | | |
| 52/406 | | 16396 | GROSE | E.C. | ← |
| | | | | | ← Unit not listed |

| Canadian/ London Gazette issue No. | Canadian/ London Gazette date | Personal Service number | Names | Initials | Remarks |
|---|---|---|---|---|---|
| **ROYAL CANADIAN AIR FORCE** | | | | | |
| **FOR EITHER NON-OPERATIONAL SERVICES IN JAPAN IN CONNECTION WITH UNITED NATIONS OPERATIONS DURING KOREAN WAR OR FOR SERVICES IN THE KOREAN AIRLIFT (*)** | | | | | |
| **QUEEN'S COMMENDATION FOR VALUABLE SERVICES IN THE AIR BRONZE OAK LEAF EMBLEM** | | | | | |
| **FLIGHT LIEUTENANT** | | | | | |
| not listed | | 27144 | DALTON | J.T. | ← Unit not listed |
| **FLYING OFFICERS** | | | | | |
| not listed | | 45258 | HATTON | B.G. | ← |
| not listed | | 28017 | SPICER | C.C. | ← |
| not listed | | 2570 | TEMPLE | L.D. | ← |
| **LEADING AIRCRAFTMAN** | | | | | |
| Not listed | | 26852 | MALO | J.W. | ← |
| | | | | | ← Unit not listed |

(*) official listings do not determine whether above awards made in recognition of services connected with United Nations operations during Korean War or for services in the Korean Airlift. These placed in this category at the Editor's discretion.

| London Gazette issue No. | London Gazette date | Personal Service number | Name | Initials | Remarks |
|---|---|---|---|---|---|

ROYAL AUSTRALIAN NAVY

OBE: OFFICER OF THE ORDER OF THE BRITISH EMPIRE

COMMANDERS

| | | | | | |
|---|---|---|---|---|---|
| | | not listed | GELLATLY | L. | HMAS Commonwealth |
| | | not listed | STEPHENSON | C.J. | ← Station not listed |

MBE: MEMBER OF THE ORDER OF THE BRITISH EMPIRE

LIEUTENANT-COMMANDER (S)

| | | | | | |
|---|---|---|---|---|---|
| | | not listed | MILLER | A.L. | ← Station not listed |

| London Gazette issue No. | London Gazette date | Personal Service number | Name | Christian names | Remarks |
|---|---|---|---|---|---|
| ROYAL AUSTRALIAN ARMY | | | | | |
| HEADQUARTERS/STAFF | | | | | |
| CBE: COMMANDER OF THE ORDER OF THE BRITISH EMPIRE | | | | | |
| BRIGADIER | | | | | |
| 40058 | 29.12.53 | 3/32 | CAMPBELL, DSO | Ian Ross | ← Command not listed |
| OBE: OFFICER OF THE ORDER OF THE BRITISH EMPIRE | | | | | |
| LIEUTENANT-COLONELS | | | | | |
| 40058 | 29.12.53 | not listed | HARRISON, ED | Clarence Stanley | Australian Army Labour Service |
| 39871 | 01.06.53 | 3/37502 | KING, ED | Lesley Dudley | Staff Corps: BCFK |
| MBE: MEMBER OF THE ORDER OF THE BRITISH EMPIRE | | | | | |
| MAJORS | | | | | |
| | | VX/700235 | CROSBY | W.J. | Staff Corps: BCFK |
| 40058 | 29.12.53 | 1/7512 | EAST | Colin Hubert Alan | Staff Corps |

| London Gazette issue No. | London Gazette date | Personal Service number | Name | Christian names | Remarks |
|---|---|---|---|---|---|
| ROYAL AUSTRALIAN CORPS OF ENGINEERS | | | | | |
| MBE: MEMBER OF THE ORDER OF THE BRITISH EMPIRE | | | | | |
| LIEUTENANT | | | | | |
| 39738 | 30.12.52 | WX33689 | ALFORD | Eric George | BCOF |
| WARRANT OFFICER CLASS I | | | | | |
| 39891 | 01.06.53 | QX61204 | LAUNDER | Charles Edward | BCOF |
| BEM: BRITISH EMPIRE MEDAL | | | | | |
| WARRANT OFFICER CLASS II | | | | | |
| | | 1/983 | SCHOLL | C. | BCOF |
| STAFF SERGEANT | | | | | |
| 39560 | 30.05.52 | NX501152 | GORDON | Allan David | BCOF |
| SERGEANTS | | | | | |
| 40058 | 29.12.53 | 2/3620 | HINES | Richard Arthur | BCOF |
| 39871 | 01.06.53 | NX171342 | MURPHY | Leonard Roy | BCOF |

| London Gazette issue No. | London Gazette date | Personal Service number | Name | Christian names | Remarks |
|---|---|---|---|---|---|
| ROYAL AUSTRALIAN CORPS OF SIGNALS | | | | | |
| OBE: OFFICER OF THE ORDER OF THE BRITISH EMPIRE | | | | | |
| LIEUTENANT-COLONEL | | | | | |
| 39738 | 30.12.52 | 3/74 | GREVILLE | Sydney Jamieson | BCOF |
| MBE: MEMBER OF THE ORDER OF THE BRITISH EMPIRE | | | | | |
| CAPTAINS | | | | | |
| 39486 | 07.03.52 | QX48209 | CHRISTIE, ED | John | BCOF |
| 40058 | 29.12.53 | not listed | KING, ED | Gordon Charles Allen | BCOF |

| London Gazette issue No. | London Gazette date | Personal Service number | Name | Christian names | Remarks |
|---|---|---|---|---|---|
| ROYAL AUSTRALIAN INFANTRY CORPS | | | | | |
| MBE: MEMBER OF THE ORDER OF THE BRITISH EMPIRE | | | | | |
| LIEUTENANT | | | | | |
| 39560 | 30.05.52 | 2/400340 | BEACROFT | Edward Copeland | 1st RHU |
| BEM: BRITISH EMPIRE MEDAL | | | | | |
| STAFF SERGEANTS | | | | | |
| 39871 | 01.06.53 | 1/400039 | REES | Thomas Robert | 1st RHU (also listed as Sgt) |
| 39560 | 30.05.52 | 2/1131 | SAINSBURY | John Henry | BCOF |
| SERGEANTS | | | | | |
| 39738 | 01.01.53 | 6/303 | BARNES | Clement Maxwell | BCOF |
| 40058 | 29.12.53 | 2/3942 | CORCORAN | Robert Brian | ← Unit not listed |
| 39560 | 30.05.52 | 1/921 | DEED | Raymond | 1st RHU |
| 39486 | 07.03.52 | 3/2295 | DICKSON | David Charles Sinclair | BCOF |

| London Gazette issue No. | London Gazette date | Personal Service number | Name | Christian names | Remarks |
|---|---|---|---|---|---|
| ROYAL AUSTRALIAN ARMY SERVICE CORPS | | | | | |
| MBE: MEMBER OF THE ORDER OF THE BRITISH EMPIRE | | | | | |
| CAPTAIN | | | | | |
| 39891 | 19.06.53 | 2/37662 | RIORDAN | Michael John | ← Unit not listed |
| BEM: BRITISH EMPIRE MEDAL | | | | | |
| WARRANT OFFICER CLASS II | | | | | |
| 39871 | 01.06.53 | 1/9127 | NEILSON | Thomas | ← Unit not listed (also shown as NEILSEN) |
| 39871 | 01.06.53 | QX31732 | POOLE | Norman Charles Roy | ← Unit not listed |

| London Gazette issue No. | London Gazette date | Personal Service number | Name | Christian names | Remarks |
|---|---|---|---|---|---|
| ROYAL AUSTRALIAN ARMY MEDICAL CORPS | | | | | |
| OBE: OFFICER OF THE ORDER OF THE BRITISH EMPIRE | | | | | |
| COLONEL | | | | | |
| 39846 | 07.03.52 | VX14663 | NYE, ED | Charles Watson | BCOF |
| MBE: MEMBER OF THE ORDER OF THE BRITISH EMPIRE | | | | | |
| MAJOR | | | | | |
| 39486 | 07.03.52 | VX102730 | LLOYD | John Wesley Francis | BCOF |
| BEM: BRITISH EMPIRE MEDAL | | | | | |
| WARRANT OFFICER CLASS II | | | | | |
| 40058 | 29.12.53 | NX9383 | KELLY | George John Alfred | BCOF |

| London Gazette issue No. | London Gazette date | Personal Service number | Name | Christian names | Remarks |
|---|---|---|---|---|---|
| ROYAL AUSTRALIAN ARMY ORDNANCE CORPS | | | | | |
| OBE: OFFICER OF THE ORDER OF THE BRITISH EMPIRE | | | | | |
| MAJOR | | | | | |
| 39560 | 30.05.52 | NX76161 | SMITH | Albert Arthur | ← Unit not listed |

| London Gazette issue No. | London Gazette date | Personal Service number | Name | Christian names | Remarks |
|---|---|---|---|---|---|

ROYAL AUSTRALIAN CORPS OF ELECTRICAL AND MECHANICAL ENGINEERS

BEM: BRITISH EMPIRE MEDAL

STAFF SERGEANT

| | | | | | |
|---|---|---|---|---|---|
| 39738 | 01.01.53 | 1/9090 | BUTLER | Victor | BCOF |

CRAFTSMAN

| | | | | | |
|---|---|---|---|---|---|
| 39738 | 01.01.53 | 2/2790 | RUSSELL | Robert James | BCOF<br>(also listed as a Warrant Officer) |

| London Gazette issue No. | London Gazette date | Personal Service number | Name | Christian name | Remarks |
|---|---|---|---|---|---|

ROYAL AUSTRALIAN ARMY PAY CORPS

MBE: MEMBER OF THE ORDER OF THE BRITISH EMPIRE

CAPTAIN

| | | | | | |
|---|---|---|---|---|---|
| 39891 | 19.06.53 | NX130710 | CAREY | Maurice | ← Unit not listed |

| London Gazette issue No. | London Gazette date | Personal Service number | Name | Christian names | Remarks |
|---|---|---|---|---|---|
| ROYAL AUSTRALIAN ARMY NURSING CORPS | | | | | |
| RRC: ADDITIONAL MEMBERS OF THE ROYAL RED CROSS, CLASS II | | | | | |
| MAJOR | | | | | |
| 40058 | 29.12.53 | F3/131 | THOMPSON | Dulcie Vera | British Commonwealth General Hospital |
| CAPTAINS | | | | | |
| | | F2/2 | McCARTHY | P.M. | BCOF |
| 39871 | 26.05.53 | F3/3 | WILDING | Helene Joyce | BCOF |
| BEM: BRITISH EMPIRE MEDAL | | | | | |
| STAFF SERGEANT | | | | | |
| 39486 | 07.03.52 | NFX66810 | HOWELL | Edna Edith | BCOF |
| SERGEANT | | | | | |
| 40058 | 29.12.53 | QFX700098 | BAKER | Eileen Maud | BCOF |

| London Gazette issue No. | London Gazette date | Personal Service number | Name | Christian names | Remarks |
|---|---|---|---|---|---|
| ROYAL AUSTRALIAN AIR FORCE | | | | | |
| OBE: OFFICER OF THE ORDER OF THE BRITISH EMPIRE | | | | | |
| WING COMMANDER | | | | | |
| 39891 | 19.06.53 | 04398 | MORGAN | David Archibald Stevenson | 391 Squadron |
| SQUADRON LEADER | | | | | |
| 39560 | 30.05.52 | 03590 | LEOPOLD | Carl Joseph | € Unit not listed |
| MBE: MEMBER OF THE ORDER OF THE BRITISH EMPIRE | | | | | |
| SQUADRON LEADER | | | | | |
| 39738 | 30.12.52 | 03193 | HARDY | William John | € Unit not listed |
| FLIGHT LIEUTENANT | | | | | |
| 39891 | 19.06.53 | 031846 | PROUD | Anthony Trevor | 391 Squadron |
| WARRANT OFFICER | | | | | |
| 40058 | 29.12.53 | A2320 | NEGUS | John Alfred Charles | 391 Squadron |
| BEM: BRITISH EMPIRE MEDAL | | | | | |
| SERGEANT | | | | | |
| 39560 | 30.05.52 | A486 | ROBERTS | Alexander Charles | 391 Squadron (number also shown as A465) |

| London Gazette issue No. | London Gazette date | Personal Service number | Name | Christian names | Remarks |
|---|---|---|---|---|---|
| ROYAL AUSTRALIAN AIR FORCE NURSING SERVICE | | | | | |
| RRC: ADDITIONAL MEMBER OF THE ROYAL RED CROSS, CLASS II | | | | | |
| SENIOR SISTER | | | | | |
| 39891 | 19.06.53 | N33303 | SCHOLZ | Phillis Reta | att 391 Squadron |

| London Gazette issue No. | London Gazette date | Personal Service number | Name | Christian names | Remarks |
|---|---|---|---|---|---|
| NON-OPERATIONAL SERVICES IN JAPAN IN CONNECTION WITH UNITED NATIONS OPERATION DURING KOREAN WAR | | | | | |
| ROYAL NEW ZEALAND ARTILLERY | | | | | |
| BEM: BRITISH EMPIRE MEDAL | | | | | |
| STAFF SERGEANT | | | | | |
| 40058 | 29.12.53 | 203880 | SMITH | Ralph Mitchell | ✱ Unit not listed |

| London Gazette issue No. | London Gazette date | Personal Service number | Name | Christian names | Remarks |
|---|---|---|---|---|---|
| ROYAL NEW ZEALAND INFANTRY | | | | | |
| MBE: MEMBER OF THE ORDER OF THE BRITISH EMPIRE | | | | | |
| CAPTAIN | | | | | |
| 39871 | 01.06.52 | not listed | LLEWELLYN | Stephen Peter | ← Unit not listed |

| London Gazette issue No. | London Gazette date | Personal Service number | Name | Christian names | Remarks |
|---|---|---|---|---|---|

ROYAL NEW ZEALAND ARMY PAY CORPS

MBE: MEMBER OF THE ORDER OF THE BRITISH EMPIRE

CAPTAIN

| London Gazette issue No. | London Gazette date | Personal Service number | Name | Christian names | Remarks |
|---|---|---|---|---|---|
| 39738 | 01.01.53 | 206376 | MALLASCH | Trevor John William | ≮ Unit not listed |

AWARDS

FOR

SERVICE ON THE KOREAN AIRLIFT

| London Gazette issue No. | London Gazette date | Personal Service number | Name | Christian names | Remarks |
|---|---|---|---|---|---|
| ROYAL CANADIAN AIR FORCE | | | | | |
| OBE: OFFICER OF THE ORDER OF THE BRITISH EMPIRE | | | | | |
| WING COMMANDER | | | | | |
| 39560 | 30.05.52 | 19523 | MUSSELLS, DSO, DFC | Campbell Hailburton | ← Unit not listed |
| MBE: MEMBER OF THE ORDER OF THE BRITISH EMPIRE | | | | | |
| SQUADRON LEADER | | | | | |
| 39560 | 30.05.52 | 1874 | LORD | William Henry | ← |
| AFC: AIR FORCE CROSS | | | | | |
| WING COMMANDER | | | | | |
| 39560 | 30.05.52 | 19984 | MORRISON, DSO, DFC | Howard Allan | ← |
| SQUADRON LEADER | | | | | |
| 39560 | 30.05.52 | 19812 | DICKSON, DFC, DFM | James Donald | ← |
| FLIGHT LIEUTENANT | | | | | |
| 39738 | 01.01.53 | 30018 | EDWARDS | Robert Martin | ← |
| FLYING OFFICER | | | | | |
| 39560 | 30.05.52 | 26365 | PAYNE, DFC | Donald Melvin | ← |
| AFM: AIR FORCE MEDAL | | | | | |
| FLIGHT SERGEANT | | | | | |
| 39560 | 30.05.52 | 21603 | DRACKLEY | Alfred Arthur | ← |
| BEM: BRITISH EMPIRE MEDAL | | | | | |
| FLIGHT SERGEANT | | | | | |
| 39560 | 30.05.52 | 2414 | ENGLEBERT, CD | Arthur Le Roy | ← |
| CORPORAL | | | | | |
| 39738 | 01.01.53 | 24068 | TRUDEL | Jean Baptiste Paul Aime | ← |
| QUEEN'S COMMENDATION FOR VALUABLE SERVICES IN THE AIR | | | | | |
| BRONZE OAK LEAF EMBLEM | | | | | |
| FLIGHT LIEUTENANTS | | 17486 | BURN | R.E. | ← |
| | | 20368 | PEARCE | D.R. | ← |
| | | 25486 | WOLKOWSKI | B.R. | ← |

← Unit not listed

| London Gazette issue No. | London Gazette date | Personal Service number | Name | Christian names | Remarks |
|---|---|---|---|---|---|
| ROYAL CANADIAN AIR FORCE | | | | | |
| QUEEN'S COMMENDATION FOR VALUABLE SERVICES IN THE AIR (continued) | | | | | |
| BRONZE OAK LEAF EMBLEM | | | | | |
| FLYING OFFICER | | | | | |
| | | 25597 | WILSON | J.P. | ← Unit not listed |
| SERGEANTS | | | | | |
| | | 25880 | HOWARD | G. | ← |
| | | 23347 | POTEKAL | L.C. | ← |
| | | | | | ← Unit not listed |

# BIBLIOGRAPHY

London Gazette

Official Korean War award listings:

Lords Commissioners of the Admiralty

Canadian National Defence HQ, Directorate of History

Australian National Defence HQ, Directorate of History

New Zealand National Defence HQ, Directorate of History

Indian National Defence HQ, Directorate of History

South African National Museum of Military History, Curator: Medals Director

"Korea: Canada's Forgotten War" by J. Melady

"Strange Battleground" by Lt-Col. Herbert Fairlie Wood, Canadian National Defence HQ

"Canadian Naval Operations in Korean Waters 1950-55" by Thor Thorgrimmson and E.C. Russell, Canadian Forces HQ, Department of National Defence, Ottawa, 1965

"Tracing the Regiments" by Captain Edgar Letts (late The Royal Warwickshire Regiment and The Royal Regiment of Fusiliers)

"Thirty Eighth Parallel" by Peter Gaston

"The Imjin Roll" by Colonel E.D. Harding, DSO

"41 Independent Commando, Royal Marines, Korea 1950-52", published by The Royal Marines Historical Society

"The Die-Hards in Korea" by Colonel J.N. Shipster, CBE, DSO

"The Argylls in Korea" by Lt-Col. G.I. Malcolm of Poltalloch

"The Borderers in Korea" by Major General J.F.M. MacDonald, CB, DSO, OBE

"The Royal Ulster Rifles in Korea" by Hugh Hamill

"Sometimes Forgotten" by Frederick Kirkland, OAM, JP

"Medals and Decorations" by Ian Angus

"British Forces in the Korean War" edited by Ashley Cunningham-Boothe and Peter Farrar

For their contribution:

Dr. Peter Farrar, historian and Editor BKVA, former Royal Fusiliers (City of London Regiment)
The late Alan A.J. Moody, Chairman, The British Korean Veterans Association, former Middlesex Regiment and RCMP
Clyde R. Bougie, CD, of Barrie, Ontario, Canada, Founding-President, Korean War Veterans Association of Canada
Leslie Peate of Ottawa, Canada, British and Canadian Korean War Veterans Associations
Captain John Clark, MC (ret'd.) of London, Canada, past National-President, Korean War Veterans Association, Canada
Derek Kinne, GC, now of Arizona, USA, former Royal Northumberland Fusiliers
Major David Sharp, BEM, BKVA, Ashford Kent, former Royal Northumberland Fusiliers
Major Francis S.G. Shore, MC, former 45 Field Regiment, Royal Artillery and BKVA
Major Kenneth U. Fraser, HQ King's Own Scottish Borderers, Berwick-upon-Tweed, and BKVA
George Hodkinson, DCM, BKVA, London, former Royal Fusiliers (City of London Regiment)
George A. Clark, MM, BKVA, Birmingham, former Royal Fusiliers (City of London Regiment)
Lieutenant-Commander John F. Cooke, MBE, DSM, RN (ret'd.), BKVA, Saltash, Cornwall, former HMS Cockade
Roy Neep, RN (ret'd.), Technical Director of Dollar Helicopters, Coventry, former HMS Glory: 801 Sqdn. Fleet Air Arm
Peter G. Warde, RN (ret'd.) of Olney, Buckinghamshire, former HMS Glory
Ronald Capers, RN (ret'd.), Chairman of the Royal Naval Association, Royal Leamington Spa
Alfred W. Buckingham, Royal Warwickshire Regimental Museum, St John's House, Warwick
The Director and Staff, Central Reference Library, City of Birmingham
Judith Machin, Deputy District Head Librarian (Warwickshire), Central Library, Royal Leamington Spa
John T. Mock, Secretary and Editor, Naval Historical Collectors and Research Association, Portishead, Bristol
Mota SINGH of Royal Leamington Spa and Warwickshire County Councillor
Jagjit Kaur KOHLI, Warwickshire County Library Services, Royal Leamington Spa, former Indian Army
Daniel German and Shirley Neill, researchers, Directorate of History, National Defence HQ, Ottawa, Canada
Brigadier H.M. Khanna, Military Adviser,The High Commission of India, India House, London
I.D. McGilchrist, New Zealand Defence Staff, New Zealand House, London
The Defence Attache, South African Embassy, London

and for their encouragement:

Writer and poet, Ian E. Kaye of Tain, Ross-Shire, former Black Watch and Argyll and Sutherland Highlanders, and BKVA
T. Patrick Haley of Banbury, Oxfordshire, former King's Own Scottish Borderers
Dennis W. Green, Chairman and Vice President, Birmingham No.1 Branch, BKVA, former Royal Leicestershire Regiment
Andrea Allum of Kenilworth, Warwickshire

and special thanks for their major contribution to:

Philip and Sarah Garrett of Radford Semele, Royal Leamington Spa, Warwickshire

# CONTENT

| | |
|---|---|
| 131 - 132 | Military: Headquarters (Staff) awards |
| 133 - 135 | Royal Canadian Horse Artillery awards |
| 136 - 137 | Lord Strathcona's Horse awards |
| 138 - 139 | Royal Canadian Artillery awards |
| 140 - 141 | Royal Canadian Corps of Engineers awards |
| 142 - 143 | Royal Canadian Corps of Signals awards |
| 144 - 147 | The Royal Canadian Regiment awards |
| 148 - 151 | The Royal 22e Regiment awards |
| 152 - 155 | Princess Patricia's Canadian Light Infantry awards |
| 156 | Royal Canadian Army Chaplains' Department awards |
| 157 - 158 | Royal Canadian Army Service Corps awards |
| 159 - 160 | Royal Canadian Army Medical Corps awards |
| 161 | Royal Canadian Army Ordnance Corps awards |
| 162 | Royal Canadian Army Corps of Electrical and Mechanical Engineers awards |
| 163 | Royal Canadian Army Pay Corps awards |
| 164 | Canadian Military Provost Corps awards |
| 165 | Royal Canadian Army Dental Corps awards |
| 166 | Canadian Army Intelligence Corps awards |
| 167 | Canadian Postal Corps awards |
| 168 - 170 | Royal Canadian Air Force awards |
| 171 - 172 | Non-combatant supporting services to Canadian Armed Forces, Korea |
| 171 | YMCA award |
| 172 | Salvation Army award |
| 173 - 176 | Royal Australian Navy awards |
| 177 - 179 | Military: Headquarters (Staff) awards |
| 180 | Royal Australian Artillery awards |
| 181 | Royal Australian Corps of Engineers awards |
| 182 - 187 | The Royal Australian Regiment awards |
| 188 | Royal Australian Army Chaplains' Department awards |
| 189 | Royal Australian Army Service Corps awards |
| 190 | Royal Australian Army Medical Corps awards |
| 191 | Royal Australian Army Ordnance Corps awards |
| 192 | Royal Australian Army Corps of Electrical and Mechanical Engineers awards |
| 193 | Australian Military Provost Corps awards |
| 194 | Australian Army Intelligence Corps awards |
| 195 | Royal Australian Army Catering Corps award |
| 196 - 210 | Royal Australian Air Force awards |

| | |
|---|---|
| 211 - 212 | Royal New Zealand Navy awards |
| 213 | Military - Headquarters (Staff) awards |
| 214 - 216 | Royal New Zealand Artillery awards |
| 217 | Royal New Zealand Corps of Engineers awards |
| 218 | Royal New Zealand Corps of Signals awards |
| 219 | The Royal New Zealand Regiment awards |
| 220 | Royal New Zealand Army Service Corps awards |
| 221 | Royal New Zealand Army Medical Corps awards |
| 222 | Royal New Zealand Army Ordnance Corps awards |
| 223 | Royal New Zealand Army Corps of Electrical and Mechanical Engineers award |
| 224 | Royal New Zealand Army Dental Corps award |
| 225 - 227 | Indian Army awards |
| 228 - 229 | South African Army awards |
| 230 - 251 | Commonwealth awards for non-operational services in Japan in connection with United Nations during the Korean War: |
| 230 - 235 | Royal Canadian Armed Forces awards |
| 236 - 248 | Royal Australian Armed Forces awards |
| 249 - 251 | Royal New Zealand Armed Forces awards |
| 252 - 253 | Royal Canadian Air Force awards for the Korean Airlift |